Tobias Schürmann

Flexible Automotive Power Management

A Predictive Auction-Based Approach

Karlsruher Beiträge zur
Regelungs- und Steuerungstechnik
Karlsruher Institut für Technologie

Band 28

Flexible Automotive Power Management

A Predictive Auction-Based Approach

by
Tobias Schürmann

KIT Scientific Publishing

Karlsruher Institut für Technologie
Institut für Regelungs- und Steuerungssysteme

Flexible Automotive Power Management:
A Predictive Auction-Based Approach

Zur Erlangung des akademischen Grades eines Doktor-Ingenieurs von der KIT-Fakultät für Elektrotechnik und Informationstechnik des Karlsruher Instituts für Technologie (KIT) genehmigte Dissertation

von Tobias Schürmann, M.Sc.

Tag der mündlichen Prüfung: 16. Juli 2024
Hauptreferent: Prof. Dr.-Ing. Sören Hohmann
Korreferent: Prof. Dr.-Ing. Eric Sax

Impressum

KIT Scientific Publishing

Karlsruher Institut für Technologie (KIT)
KIT Scientific Publishing
Straße am Forum 2
D-76131 Karlsruhe

KIT Scientific Publishing is a registered trademark
of Karlsruhe Institute of Technology.
Reprint using the book cover is not allowed.

www.bibliothek.kit.edu/ksp.php | E-Mail: info@ksp.kit.edu | Shop: www.ksp.kit.edu

Print on Demand 2025 – Gedruckt auf FSC-zertifiziertem Papier

ISSN 2511-6312
ISBN 978-3-7315-1454-1
DOI 10.5445/KSP/1000184711

Preface

This thesis results from my work as a research assistant at the FZI Research Center for Information Technology in Karlsruhe. Working there was and still is interesting and varied. Many people have accompanied and supported me during the last years of work leading to this thesis. I am very grateful for their guidance, support, and confirmation.

First and foremost, I thank my supervisor, Prof. Dr.-Ing. Sören Hohmann, for providing the scientific environment and profoundly supervising my research. Thank you for your patience and steadily reminding me of my research besides the projects at FZI. I also thank my co-supervisor, Prof. Dr.-Ing. Eric Sax, for the assessment and constructive review.

Many thanks go to the colleagues at FZI for the warm working atmosphere and to the colleagues at the Institute of Control Systems (IRS) at the Karlsruhe Institute of Technology (KIT) for the research exchange. I enjoyed my time at FZI, and even the more challenging times felt less burdensome with your support. In particular, I am grateful for my colleagues at the Department of Control in Information Technology (CIT), with whom I spent most of the time in the last eight years. I want to acknowledge and thank Dr.-Ing. Stefan Schwab, Dr.-Ing. Bálint Varga, Thomas Rudolf, Lukas Köhrer, Nina Majer, Max Grobbel, Daniel Flögel, Lars Fischer, and Jacqueline Henle for proofreading this thesis and providing valuable and honest feedback. I also thank all the students I supervised during their bachelor's or master's theses for supporting my research project and work at FZI.

At FZI, I was honored to participate in various projects. The projects with Mercedes-Benz, in particular, inspired and enabled this thesis. Finally, the research leading to this thesis was funded by the public project KI4BoardNet and the Federal Ministry of Education and Research (BMBF), Germany, under grant No. 16ME0778.

Last but not least, I am grateful for the unconditional support of my friends and family, not least because of the regular amusing question of whether I would finish my thesis rather in years or decades. Notably, the last months of writing were only possible with your help and encouragement.

Karlsruhe, March 2025

I have nothing to offer but blood, toil, tears, and sweat. - Winston Churchill

Kurzfassung

Aufgrund der fortschreitenden Elektrifizierung, der Verbreitung von Unterhaltungs- und Informationssystemen und der wachsenden Bedeutung von autonomen Fahrfunktionen steigt die Relevanz des Fahrzeugbordnetzes stetig. Zur gleichen Zeit fordern die kurzen Updatezyklen von Unterhaltungs- und Informationssystemen und die wachsende Anzahl von Fahrzeugvarianten eine reibungslose Integration neuer Komponenten in das Fahrzeugbordnetz, um dadurch eine nachhaltige Fahrzeugplattform zu fördern. Hinsichtlich des Softwarebereichs verhindern vordefinierte und zentralisierte Managementstrukturen und eine signalbasierte Kommunikation diese Plug-and-Play Integration, da jede Änderung im Hardwarebereich einen hohen Aufwand für die Neuauslegung des Leistungsmanagements und der Kommunikation erfordert.

Deshalb wird in dieser Arbeit ein auktionsbasiertes Leistungsmanagement für das Automobil untersucht, formalisiert und erweitert, um diese Herausforderung anzugehen. Zuerst werden aktuelle Ansätze und Konzepte im Stand der Technik, unterteilt in regel-, optimierungs-, agenten- und auktionsbasierte Ansätze, im Hinblick auf wünschenswerte Eigenschaften analysiert und qualitativ evaluiert. Mit Bezug zu den gewünschten Eigenschaften wird ein auktionsbasierter Mechanismus für moderne Fahrzeuge mit mehreren Spannungsniveaus formuliert. Um eine effektive Leistungsverteilung zwischen allen Komponenten zu gewährleisten, nutzt der Ansatz eine Auktion mit einheitlicher Preisgestaltung. Darüber hinaus wird der Ansatz mit einem flexiblen Kommunikationskonzept basierend auf einer serviceorientierten Architektur kombiniert. Zur weiteren Verbesserung wird eine prädiktive Erweiterung vorgeschlagen, die den auktionsbasierten Mechanismus komplementiert. Die prädiktive Erweiterung nutzt flexible Komfortkomponenten zur Lastverschiebung und reduziert dadurch Leistungsspitzen und Batteriebelastung. Die Effektivität wird simulativ für exemplarische Fahrszenarien gezeigt, in denen das Leistungsmanagement gebraucht wird, um die Leistungsnachfrage und -erzeugung anzugleichen. Zusammenfassend stellt das vorgeschlagene Verfahren den effektiven Leistungsausgleich in Fahrzeugen sicher und ermöglicht eine Plug-and-Play Integration neuer Hardwarekomponenten während der Entwicklung und nach der Produktion in das Fahrzeug, ohne dass eine Neuauslegung des Leistungsmanagements notwendig ist.

Kurzfassung

Aufgrund der fortschreitenden Elektrifizierung, der Verbreitung von Unterhaltungs- und Informationssystemen und der wachsenden Bedeutung von autonomen Fahrfunktionen steigt die Relevanz des Fahrzeugbordnetzes stetig. Zur gleichen Zeit fordern die kurzen Updatezyklen von Unterhaltungs- und Informationssystemen und die wachsende Anzahl von Fahrzeugvarianten eine reibungslose Integration neuer Komponenten in das Fahrzeugbordnetz, um dadurch eine nachhaltige Fahrzeugplattform zu fördern. Hinsichtlich des Softwarebereichs verhindern vordefinierte und zentralisierte Managementstrukturen und eine signalbasierte Kommunikation diese Plug-and-Play Integration, da jede Änderung im Hardwarebereich einen hohen Aufwand für die Neuauslegung des Leistungsmanagements und der Kommunikation erfordert.

Deshalb wird in dieser Arbeit ein auktionsbasiertes Leistungsmanagement für das Automobil untersucht, formalisiert und erweitert, um diese Herausforderung anzugehen. Zuerst werden aktuelle Ansätze und Konzepte im Stand der Technik, unterteilt in regel-, optimierungs-, agenten- und auktionsbasierte Ansätze, im Hinblick auf wünschenswerte Eigenschaften analysiert und qualitativ evaluiert. Mit Bezug zu den gewünschten Eigenschaften wird ein auktionsbasierter Mechanismus für moderne Fahrzeuge mit mehreren Spannungsniveaus formuliert. Um eine effektive Leistungsverteilung zwischen allen Komponenten zu gewährleisten, nutzt der Ansatz eine Auktion mit einheitlicher Preisgestaltung. Darüber hinaus wird der Ansatz mit einem flexiblen Kommunikationskonzept basierend auf einer serviceorientierten Architektur kombiniert. Zur weiteren Verbesserung wird eine prädiktive Erweiterung vorgeschlagen, die den auktionsbasierten Mechanismus komplementiert. Die prädiktive Erweiterung nutzt flexible Komfortkomponenten zur Lastverschiebung und reduziert dadurch Leistungsspitzen und Batteriebelastung. Die Effektivität wird simulativ für exemplarische Fahrszenarien gezeigt, in denen das Leistungsmanagement gebraucht wird, um die Leistungsnachfrage und -erzeugung anzugleichen. Zusammenfassend stellt das vorgeschlagene Verfahren den effektiven Leistungsausgleich in Fahrzeugen sicher und ermöglicht eine Plug-and-Play Integration neuer Hardwarekomponenten während der Entwicklung und nach der Produktion in das Fahrzeug, ohne dass eine Neuauslegung des Leistungsmanagements notwendig ist.

Abstract

Due to the ongoing electrification, the dissemination of entertainment and infotainment systems, and the growing importance of autonomous driving functions, the relevance of the automotive power network is steadily increasing. At the same time, the short update cycles for entertainment and infotainment hardware and the growing number of vehicle variants require seamless integration of new components into the power network to encourage sustainable vehicle platforms. Concerning the software domain, predefined and centralized management structures and signal-based communication of present vehicle platforms impair the plug-and-play integration since every change in the hardware domain requires high efforts in redesigning the power management and communication.

Therefore, in this work, an auction-based automotive power management is investigated, formalized, and extended to tackle these challenges. First, state-of-the-art approaches and concepts, such as rule-based, dynamic optimization-based, agent-based, and auction-based, are analyzed and qualitatively evaluated regarding desirable properties. With respect to the desired properties, an auction-based mechanism for modern vehicles with multiple voltage levels is formulated. The approach deploys a uniform price auction to achieve effective power distribution between all components. Furthermore, the approach is combined with a flexible communication design based on a service-oriented architecture. For further enhancement, a predictive extension complementing the auction-based mechanism is proposed. The predictive extension exploits flexible comfort loads by load shifting and thereby mitigates power peaks and battery strain. The effectiveness is simulatively demonstrated for exemplary driving scenarios in which power management is needed to balance power demand and supply. In summary, the proposed approach ensures effective power balancing in vehicles and facilitates plug-and-play integration of new hardware components into the vehicle during development and after production without a power management redesign.

Contents

List of Figures

List of Tables

Abbreviations and Symbols

Abbreviations

Abbreviation	Description
APN	Automotive Power Network
BMS	Battery Management System
CAN	Controller Area Network
CAN FD	Controller Area Network Flexible Data-Rate
DDS	Data Distribution Service
ECU	Electronic Control Unit
EM	Electric Machine
EMS	Energy Management Shell
EV	Electric Vehicle
HEV	Hybrid Electric Vehicle
HVAC	Heating, Ventilation, and Air Conditioning
ICE	Internal Combustion Engine
LIN	Local Interconnect Network
MAS	Multi-Agent System
MCP	Market Clearing Price
MOSFET	Metal-Oxide-Semiconductor Field-Effect Transistor
MOST	Media Oriented Systems Transport
MPC	Model Predictive Control
OCV	Open Circuit Voltage
PE	Power Electronics
PES	Power Electronics Shell
PMS	Power Management Shell
PMSM	Permanent-Magnetic Synchronous Machine
PPF	Price-to-Power Function
SOA	Service-Oriented Architecture
SOC	State Of Charge
SOH	State Of Health
SOME/IP	Scalable service-Oriented MiddlewarE over IP
SOME/IP-SD	SOME/IP Service Discovery
UPA	Uniform Price Auction

Symbols

Symbol	Description
$\square_{12V}$	Index for the 12V battery
$\square_{48V}$	Index for the 48V battery
A_i	Surface of the component i
α_i	Coefficient of heat exchange between the air and the surface i
$\mathcal{B}$	Group of bidders in the uniform price auction
$B_j(p)$	Bid of the bidder j according to the price p
C	Capacitance
C_{air}	Thermal capacitance of the cabin air
c_{air}	Specific thermal capacitance of the cabin air
Δp_{rec}	Price interval for increased power availability due to recuperation
Δp_{norm}	Price interval for normal APN operation
Δp_{red}	Price interval for comfort component reduction
Δp_{crit}	Price interval during critical APN status
$\Delta\dot{H}$	Difference between enthalpies
Δp_{blower}	Relative price range for the blower reduction
Δp_{seat}	Relative price range for the seat heating reduction
ΔT	Time interval
ϵ	Algorithmic threshold to calculate the MCP
ϵ_{rad}	Specific absorption of radiation
E_{solar}	Solar radiation
$\dot{H}_{\text{in}}$	Enthalpy of the incoming airflow
h_{in}	Specific enthalpy of the incoming air
$\dot{H}_{\text{out}}$	Enthalpy of the outgoing airflow
h_{out}	Specific enthalpy of the outgoing air
i	Electric current
J	Objective function of the MPC
L	Inductance
$\lambda_{\text{Batt}}(T_{\text{Batt}})$	Battery power factor depending on the battery temperature
$\lambda_{\text{Batt, c}}(T_{\text{Batt}})$	Factor for battery charging power
$\lambda_{\text{Batt, dis}}(T_{\text{Batt}})$	Factor for battery discharging power
λ_s	Parameter to shape the battery price-to-power function slope
m_{air}	Mass of the cabin air
$\text{MCP}(k)$	Market clearing price for the time interval k
M_{gen}	Negative torque of the EM in generator mode
λ_{gen}	Growth factor of the EM power generation
$M_{\text{ref, EM}}$	Desired EM torque

Symbol	Description
P_{batt}	Charging or discharging power of the battery
n_r	Rotation speed of the rotor magnetic field in rpm
n_s	Rotation speed of the stator magnetic field in rpm
N_p	Length of the prediction horizon of the MPC
$O_i(p)$	Offer of the seller i according to the price p
P_{ada}	Provided power in the adaptive power generation mode of the EM
P_{blower}	Sum of all blower stepwise switching power stages
$P_{\text{set, boost}}$	Reference power in the boosting mode of the EM
P_{boost}	Consumed power in the boosting mode of the EM
$P_{\text{c, max}}$	Maximum battery charging power
P_{c}	Actual battery charging power
$P_{\text{dis, max}}$	Maximum battery discharging power
P_{dis}	Actual battery discharging power
P_{gen}	Maximum power supply of the EM in generator mode
$\hat{P}_{\text{HVAC}}$	Estimated HVAC power consumption
$P_{\text{HVAC, max}}$	HVAC maximum power consumption
$P_{\text{HVAC, norm}}$	HVAC power consumption for the intended operation
P^{lm}	Power transfer from voltage level l to m
$P^{\text{12-to-48}}$	Power transfer from voltage level 12V to 48V
$P^{\text{48-to-12}}$	Power transfer from voltage level 48V to 12V
P^{lm}_{ref}	Power transfer commands for the DC/DC-converter
$P^{lm}_{\text{PE, lim}}$	Power transfer limitation from voltage level l to m
$P^{\text{12-to-48}}_{\text{PE, lim}}$	Power transfer limitation from voltage level 12V to 48V
$P^{\text{48-to-12}}_{\text{PE, lim}}$	Power transfer limitation from voltage level 48V to 12V
P_{low}	Lower power limit concerning the normal price range Δp_{norm}
P_{seat}	Sum of all seat heating stepwise switching power stages
P_{switch}	Sum of all stepwise switching power stages
P_{up}	Upper power limit concerning the normal price range Δp_{norm}
$\text{PPF}_{\text{cum}, m}$	Cumulative PPF of the voltage level m
$\text{PPF}_i(p)$	Price-to-power function of component i according to the price p
P_{rec}	Provided power in the recuperation mode of the EM
$P_{\text{rec, set}}$	Reference power supply in the recuperation mode of the EM
P_{ref}	Power references for the APN systems
p_{crit}	Lower threshold of the price interval Δp_{crit}
p_{init}	Initial market price for the uniform price auction
p_k	Market price in the next time instant k
p_k^m	Market price for the time instant k and the voltage level m
p_{max}	Maximum market price of the valid price range
p_{min}	Minimum market price of the valid price range
p_{norm}	Lower threshold of the price interval Δp_{norm}
p_{red}	Lower threshold of the price interval Δp_{red}
p_x	Price threshold for battery charging and discharging

Symbol	**Description**
Ψ	Magnetic flux
$\dot{Q}_{\mathrm{conv,\,in}}$	Convection heat between components and the cabin air
$\dot{Q}_{\mathrm{HVAC}}$	Heat input from the HVAC
$\dot{Q}_{\mathrm{pers}}$	Convection heat input from the persons in the vehicle
$\dot{Q}_{\mathrm{rad}}$	Heat emission by the vehicle components through radiation
$\dot{Q}_{\mathrm{solar}}$	Heat impact of the solar radiation
R_s	Resistor of the electric machine stator winding
$R_{\mathrm{uc,\,p}}$	Parallel resistor of the ultracapacitor
$R_{\mathrm{uc,\,s}}$	Serial resistor of the ultracapacitor
S_i	Switch i
s_{EM}	Slip of the asynchronous machine
$\mathcal{S}$	Group of seller in the uniform price auction
σ_{rad}	Stefan-Boltzmann constant for radiation
$\mathrm{SOC}_{\mathrm{ref}}$	SOC reference for a storage system
S_{ref}	Switching reference for a power electronics component
$\tau_{\mathrm{Batt,\,1}}$	First current-voltage time constant of the battery
$\tau_{\mathrm{Batt,\,2}}$	Second current-voltage time constant of the battery
τ_{EM}	Current-voltage time constant of the EM
τ_{uc}	Current-voltage time constant of the ultracapacitor
T	Temperature
T_{amb}	Temperature of the ambient air
T_{body}	Temperature of the vehicle body
T_{cab}	Temperature of the cabin air
$\Delta T_{\mathrm{cab,\,acc}}$	Acceptable cabin temperature range
$T_{\mathrm{cab,\,ref}}$	Reference cabin temperature set by the passengers
T_{front}	Temperature of the front windshield
T_i	Temperature of the component i
T_{int}	Temperature of the interior surfaces
T_{pers}	Temperature of the persons in the vehicle
T_{rear}	Temperature of the rear windshield
T_{side}	Temperature of the side windows
$\underline{u}_{\mathrm{power}}$	Lower power limit of the HVAC system control
$\overline{u}_{\mathrm{power}}$	Upper power limit of the HVAC system control
V_{cell}	Cell voltage of the battery
V_{OC}	Open circuit voltage of the battery
V_{uc}	Output voltage of the ultracapacitor
$\mathbb{V}$	Set of automotive power network voltage levels
V_{ref}	Voltage reference of an APN voltage level

1 Introduction

In recent years, the number of electric and electronic components in the vehicle has increased significantly with up to 100 electronic control units (ECUs) [WSD+18, MPPRM+20, KKv+21, TVE+22]. The electrification of former mechanical components, the growing number of telecommunication and entertainment systems, and the dissemination of advanced driver-assistance systems and autonomous driving functions are the three major trends leading to this development.

The electrification of former mechanical components, such as the heating, ventilation, and air conditioning (HVAC) system, the brake booster, the window lifter, or the drive train in hybrid electric vehicles (HEVs) and electric vehicles (EVs), brings advantages in terms of efficiency and control performance. At the same time, new telecommunication and entertainment systems have evolved rapidly, permeating more areas of everyday life and integrated into vehicles, making them the new living room [LH02, MPPRM+20]. As a result, complexity, caused by the growing number of components and their interaction, further increases. Moreover, there is a demand for always-available vehicles. This demand concerns software updates over the air and the availability for the customer via mobile apps [ORBE20, RGKS20]. The third trend is the dissemination of advanced driver-assistance systems and autonomous driving functions, which is still in progress. To enable autonomous driving functions, sensors for the perception task, additional computational power, and electric actuators are required, which all pose high power demands to the vehicle. All these trends add electric components and ECUs to the physical automotive power network (APN) as depicted in Figure 1.1 and emphasized by the loads with the three dots.

With the growing number of electric consumers in the APN, the average and peak power consumption in the multiple voltage levels of vehicles has significantly increased and is expected to grow further [Büc08, KWT+10, TPFH21]. As a result, modern and future APNs face similar challenges to energy grids in terms of stability and power distribution [VGM+10]. Hence, a capable power management is essential to operate the power network effectively. The automotive power management, as depicted in Figure 1.1, is connected to all components, aggregating information about the APN state and sending control requests to the active components, for instance, the electric machine (EM) or the HVAC system in case of power imbalances. Since the batteries are passive energy buffers, they only provide information about their current state, visualized by the one-sided arrow. In summary, capable power management is necessary to coordinate the power balance and facilitate voltage stabilization in the APN voltage levels, ensuring a reliable and robust vehicle operation [KET+11, RWM+13].

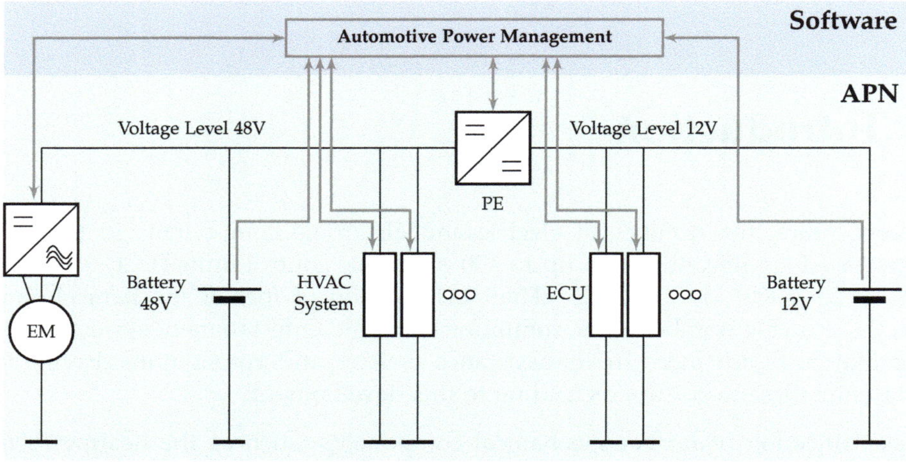

Figure 1.1: APN with two voltage levels, 12V and 48V, connected via a DC/DC-converter (PE), the HVAC system and the ECU as loads, and a battery in each voltage level. The automotive power management in the software domain monitors the components and coordinates the power flow in the physical APN. The software domain and the physical APN are elaborated in Chapter 2 and Chapter 3, respectively.

Besides the reliable vehicle operation, the keys toward business success for modern and future cars are modularity and flexibility to enable the requested customization and the shorter update and development cycles [LH02, MPPRM+20]. In the past, the APN software module, see Figure 1.1, was predefined at the start of production and partly updated in exceptional situations during a periodical inspection [RGKS20, ORBE20]. Today, over-the-air software updates after the start of production and during operation are widespread for modern vehicles since software development cycles are shorter than the development period of new vehicle hardware platforms [KOB+17, RGKS20]. Due to the updated software functions, the functionality of components and, consequently, the vehicle's power consumption may change, which again affects automotive power management. Furthermore, the importance of software for new vehicles and updates over the air change the view on the vehicle and its maintenance [SSFK16, RGKS20]. In the past, the inspection of mechanical parts dominated the periodic maintenance. In recent years, errors in the software stack have been a growing challenge [KHP18, KKv+21]. Probably, this evolution will scale up with the rising number of electric and electronic components in the APN and the rising significance of software in the vehicle [KHP18].

In addition, in the near future, the possibility for plug-and-play integration of new or updated hardware components becomes an essential feature concerning a sustainable vehicle platform, which is updated regularly with the evolving telecommunication and entertainment systems [WSD+18]. These hardware updates may happen during the development process or even after production. Nowadays, the intense customization of the automobile already leads to millions of variants, resulting in the current situation that there are rarely two cars completely equally equipped [MPPRM+20]. Concerning the development processes, this accounts for rising efforts and, on the other hand, for a growing number of software failures, for example through strongly interconnected systems and their dependencies [BAJ+13, WSD+18].

1.1 Research Goal

The current and future challenges for APNs are the higher power consumption, the shorter software update cycles, and the need for a flexible hardware composition even after manufacturing. To address these challenges, the strong interconnection between hardware and software calls for a change in both areas [WB18, p. 7f].

On the hardware side, the concept of a zonal architecture and flexible connectors in [TTFH18, BLOT19, TPFH21, ZZL+21] together with adaptable power resources may provide further flexibility for the APN. The new architectures involve the power infrastructure as well as capable and flexible communication technologies, in particular, Automotive Ethernet [HLZ16, RGKS20]. Additionally, ECUs evolve toward scalable and centralized computation platforms offering possibilities in the design process [RWKT13, WG20]. In summary, the available hardware technologies and concepts facilitate the flexible APN composition and a plug-and-play hardware integration after the production [Ema05, TPFH21]. Therefore, it is appropriate to assume the availability of a flexible APN hardware platform in this thesis.

However, on the software side, the current automotive power management approaches and the signal-based communication scheme impair the seamless plug-and-play integration of new hardware components and cause significant development efforts [WA08, RGKS20, HSS+22]. Hence, this thesis focuses on designing and developing an effective and flexible power management based on the uniform price auction (UPA) that coordinates the power consumption in the vehicle. Accordingly, in case of power imbalances, the power management interferes and balances the power by short-term load adjustments of comfort loads and supply system adjustments. In this regard, the basic idea in [Gra04a] is formalized, extended for APNs with multiple voltage levels, and complemented with a communication design based on a service-oriented architecture (SOA).

This work's research goal is to investigate and assess promising methodologies and develop a capable automotive power management approach supporting the desired plug-and-play integration property. The primary goal is to effectively coordinate all components to ensure a stable and reliable power supply. The second goal is to provide flexibility and modularity, facilitating the plug-and-play integration property of future APNs.

1.2 Structure of the Thesis

In accordance with the research goal, the thesis is structured as follows. Chapter 2 investigates the state-of-the-art power management approaches in the automotive field and neighboring domains. Firstly, the APN management structure is described to separate the power management task from the overall APN software as visualized in Figure 1.1. Afterward, the necessary properties for a capable power management algorithm are elaborated and defined. State-of-the-art approaches and methodologies in research that have not yet been applied to power management in real-world applications are presented. Finally, the chapter summarizes the various approaches and methodologies with a qualitative comparison and concludes with the identified research gap.

In Chapter 3, the fundamentals of the APN, the underlying hardware architecture, and the subordinate components are introduced. The chapter provides a general understanding of the APN as a physical system and derives the properties for the later design of the power management approach. Furthermore, the currently used and emerging communication technologies are presented to consider the communication between the different management layers and within the APN. The chapter concludes with the deduction of goals toward capable APNs.

Chapter 4 presents the proposed auction-based power management. After a short theoretical categorization, the auction-based approach is mathematically formalized and augmented for APNs with multiple voltage levels. Subsequently, the predictive extension of the auction-based power management is described. The aim of the predictive extension is the exploitation of flexible loads shifting power demands in order to mitigate short-term power peaks. Finally, the communication design based on a SOA is elaborated.

To demonstrate the working principle and the effectiveness of the auction-based power management for an APN with multiple voltage levels, Chapter 5 provides a simulative evaluation. The chapter includes the description of an exemplary APN and scenarios in which the power management is requested to regulate the power consumption in the APN, reducing the strain on the batteries and ensuring the vehicle's safe operation. The results of the presented simulation are analyzed and discussed with regard to the desired quality of power balancing and the plug-and-play integration of new hardware components.

2 Automotive Power Management: State of the Art and Research Gap

With respect to the formulated research goal, this chapter elaborates and assesses the state of the art concerning automotive power management.

Section 2.1 presents the software architecture of the APN management. Based on the classic structure, which is elaborated first, a new software architecture and the power management tasks with regard to the scope of the thesis are defined. The subsequently described desirable properties of automotive power management in Section 2.2 depict the aim to strive.

In Section 2.3, state-of-the-art power management approaches, methodologies in research, and approaches in neighboring domains are presented with an emphasis on the required properties. A qualitative comparison of the presented approaches and methodologies in Section 2.5 followed by the identified research gap in Section 2.6 concludes the chapter.

2.1 Software Architecture

The electric and electronic (E/E) architecture defines the composition and interplay of the electrical and electronic automotive system, which comprises different hardware and software components and systems as depicted in Figure 1.1. The software architecture is one part or perspective of the vehicular E/E architecture [Rei11, p. 155]. In recent years, the software architecture describing the distribution and location of software functions among the ECUs in the vehicle has changed from a distributed to a more centralized architecture [ORBE20]. Different approaches toward a centralized architecture include domain or zonal architectures.

The APN management system is one subsystem in the vehicle software controlling the APN in Figure 1.1. The regarded automotive power management is embedded into the overall APN management system. In [RL07], the authors define the terms energy management shell (EMS), power management shell (PMS), and power electronics shell (PES) to separate the different tasks. In the following, these tasks are elaborated with reference to the management pyramid shown in Figure 2.1 and first presented for the APN management in [RL07].

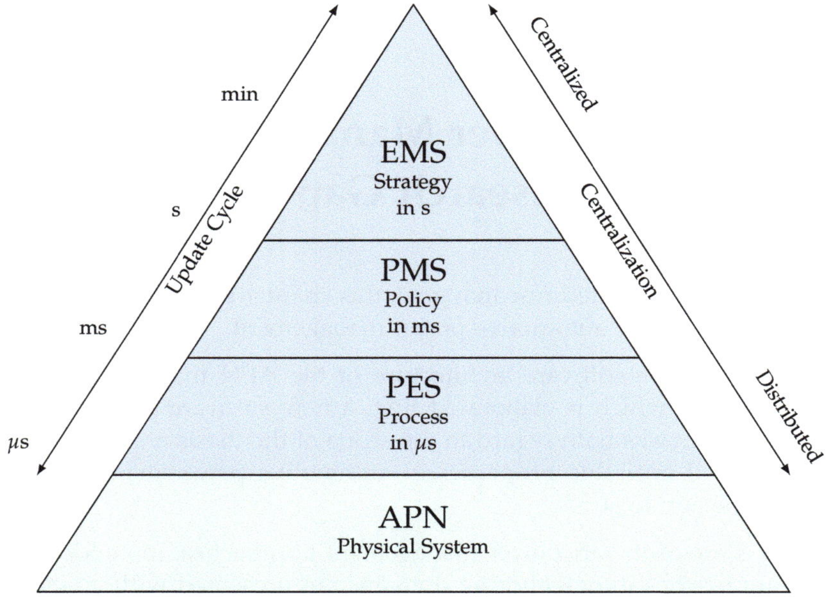

Figure 2.1: Cascaded APN management shells with decreasing update cycles to control the energy and power flow and stabilize the respective voltage levels in the APN (adapted from [RL07]). The management shells comprise the energy management shell (EMS) at the top, the power management shell (PMS) at the middle, and the power electronics shell (PES) at the bottom level, directly connected to the physical APN.

2.1.1 Classic Software Architecture and Task Classification

As depicted in [RL07] and Figure 2.1, power management is commonly viewed as one shell in a hierarchical structure. The graphical representation is derived from the general management hierarchies and is similar to the industrial automation pyramid [RL07, KBK+19]. Tasks in the energy grid are allocated in a similar hierarchy from short-term stabilization to long-term energy distribution [VGM+10, PD11].

Independent of the shells, the APN management system provides information to other functions in the vehicle, the driver, or the passengers. Hence, a general task is the aggregation and the processing of information with regard to the components and systems in the APN. Therefore, additional connections exist for information flow from the three software shells to other software functions in the vehicle that are not part of Figure 2.1.

The visualized decision chain in Figure 2.1 is a common approach to divide the tasks of energy and power management in a vehicle [RL07, EGLE18]. The specific implementation of the three shells differs for varying vehicle architectures with respect to the voltage levels, the incorporated power sources, and the power train configuration. However, the described and allocated tasks are equivalent in the automotive domain.

The management levels or shells are the EMS, the PMS, and the PES, which are all part of the APN management highlighted in blue in Figure 2.1. The underlying APN in green includes all physical components, the power consumers and suppliers, as well as the distribution networks in the electric domain of the vehicle. Accordingly, the APN shell comprises the components as well as their connections for electric power and communication. In Figure 1.1, the physical domain with various components for an HEV architecture with two voltage levels is visualized in green and marked with APN. The software level is highlighted in blue. For the description of specific relations and tasks, it is generally referred to this HEV architecture and the according components. Different HEV configurations emphasizing the hybrid propulsion system are explained in [ELR08] and [EGLE18, p. 113ff]. The APN in Figure 1.1 comprises the energy and power infrastructure, such as the harness, the DC/DC-converter, the batteries, the EM, and maybe other power sources, as well as various loads, including comfort loads, such as the HVAC system, safety-relevant loads, and computational units, for example an ECU.

Energy Management

The EMS is accountable for the coordination of electric energy during driving operation and thereby sets the long-term *Strategy* of the next time interval ranging from seconds to minutes [RLEW06, RL07, Kir19]. In this regard, the EMS controls the state of charge (SOC) of the different energy and power sources, such as the batteries or super-capacitors. From the superordinate driving strategy, the EMS receives commands or constraints regarding the propulsion, such as torque requests for the EM. Moreover, it considers information from the energy storages to determine constraints for safe and reliable operation.

In summary, the EMS

- ◇ determines reference values $\mathrm{SOC}_{\mathrm{ref}}$ and
- ◇ provides power constraints, such as P_{min} and P_{max},

for the energy storages [RL07]. The goal is to precisely implement the driving strategy and simultaneously consider the reasonable operation of the energy sources, mitigating wearing and aging effects.

As an example, for the HEV configuration in Figure 1.1, there is the possibility to divide the required driving torque for acceleration between the internal combustion engine (ICE) and the EM. Additionally, there is the recurring decision of whether to charge the battery cells or to wait for the next recuperation phase and, thus, keep free charging capacity. This leads to the reference values $\mathrm{SOC}_{\mathrm{ref},12}$ and $\mathrm{SOC}_{\mathrm{ref},48}$ for the two battery storages. Since the electric energy resource of the battery is limited, the main task is to decide whether or not to deploy the additional electric torque or to drive fully electric [KBH09]. There are comprehensive works that elaborate on this question for various vehicle architectures and mainly HEV configuration with a varying degree of hybridization [KKJ+07, EGLE18, ORBE20, RSSH21].

The EMS communication comprises two directions. On the one hand, the EMS communicates with the vehicle to determine the driving strategy. On the other hand, the EMS passes commands or reference values to the PMS and receives information about the APN. The APN information includes, in particular, the energy storage states, for instance, the current SOC, the state of health (SOH), or the temperature. This information is necessary to derive the strategy for the next time interval considering the energy storage state.

Power Management

The next shell in the decision chain toward the physical APN is the PMS, which is the focus of this thesis. Based on the provided reference values $\mathrm{SOC}_{\mathrm{ref}}$ and constraints from the EMS, the PMS determines reference values for the power supply and demand in the APN concerning the power infrastructure, such as the EM, the batteries, and the DC/DC-converter in Figure 1.1. Hence, the PMS implements the strategy from the EMS by setting the *Policy* of power utilization and distribution for the next period [RL07]. The PMS usually plans the policy for an interval of milliseconds up to seconds.

In summary, the PMS

- ◇ determines the power reference P_{ref},

which is subject to

- ◇ the provided reference values $\mathrm{SOC}_{\mathrm{ref}}$,
- ◇ power constraints for the energy storages, such as $P_{\min}$ and $P_{\max}$,
- ◇ and the desired EM torque $M_{\mathrm{ref,EM}}$.

One significant aspect of the PMS is to achieve a power balance in the APN, which is expressed by

$$\sum_{i=0}^{N} P_i = 0 \text{ for } \forall i \in \mathrm{APN}. \tag{2.1}$$

For the power balance, the power consumption and generation of all electric or electronic components i in the physical APN are aggregated. Since some components are not controlled by the APN management but depend on the passenger behavior, the power references P_{ref} have to be adapted iteratively in order to achieve the required power balance. Due to the passive nature of the battery storages, their power supply or consumption is given by

$$P_{\text{batt}} = f(V, \cdot) \tag{2.2}$$

and depends on the current voltage level V and various battery states. As a result, the PMS has to derive a certain voltage reference V_{ref} for the APN voltage levels with reference to the determined power references P_{ref} and respective voltage limits [Büc08, p. 11].

With regard to the APN in Figure 1.1, the PMS provides the reference values P_{ref} for the EM, the DC/DC-converter, and the batteries. Additionally, the voltage levels $V_{\text{ref},12}$ and $V_{\text{ref},48}$ are derived and communicated to the active components, which is the EM with its inverter in the 48V level and the DC/DC-converter concerning the 12V level. Accordingly, the PMS communicates the power references P_{ref} and the voltage references V_{ref} to the PES. In turn, the PMS receives the current power demand or supply, the current voltage levels, and additional information about the APN components.

Power Electronics

The last shell in Figure 2.1 is the PES, which is situated between the PMS and the physical components and systems in the APN. The PES includes all control cycles close to the hardware and implements the power references P_{ref} and the voltage references V_{ref} by controlling the switching signals of the power electronics (PE). Hence, the PES implements the PMS policy by adequately controlling the *Process* in the time range of microseconds to milliseconds [RL07].

In summary, the PES

- ◇ determines the switching references S_{ref}

to implement

- ◇ the provided power references P_{ref}
- ◇ the voltage references V_{ref},
- ◇ and the desired EM torque $M_{\text{ref,EM}}$ for the vehicle propulsion.

In Figure 1.1, the EM and the DC/DC-converter are the active power network components that primarily influence the respective voltage levels. Hence, the PES receives the references for power and the voltage level, P_{ref} and V_{ref}, from the PMS and translates these in switching references S_{ref} for the EM's inverter and the DC/DC-converter. From the physical APN components, the PES gets information about the current component and network states. This information is considered to derive adequate switching references S_{ref} and passed to the superordinate shells, PMS and EMS.

Control Cycles and Update Rates

Caused by the sketched control cycles and tasks for each APN management shell, the proximity to the physical components in the APN has a wide range. The EMS and the PMS, providing the strategic decisions and the derived policy, are mostly centralized software units [ORBE20]. Accordingly, they are situated in one computation unit (ECU), aggregate all information from the vehicle passed from the PES, and derive strategic decisions based on the vehicle state and information about the future driving profile [GS17, ORBE20]. Subsequently, communication structures propagate these decisions toward the PES.

In contrast, the PES as the bottom layer is commonly organized in a distributed way and located in direct proximity to the respective components [KWSK02, Büc08, KWOH17]. Since the transient behavior of the systems and, thus, the respective PES control cycles are in the range of microseconds to milliseconds, every delay in the PES communication impairs the control performance. As a result, systems with integrated PE in the vehicle need highly responsive control functions, determining the power flow [KWOH17]. Consequently, the control functions are embedded in ECUs, which are mostly located within the respective systems.

In summary, the centralization on the right side of Figure 2.1 decreases from the centralized EMS, which makes decisions for the vehicle operation, to the distributed PES, which implements the reference values on component level. Simultaneously, the update cycles on the left side decrease from a time interval of seconds to minutes for EMS decisions to microseconds for the highly responsive PE control cycles. The distribution and location of software functions, as well as the necessary communication structures and update cycles, are all part of the E/E architecture.

2.1.2 New Software Architecture

Based on the current implementations and research in terms of energy and power management algorithms, the APN management structure in Figure 2.1, which is proposed in [RL07] does not represent the state of the art. In the current research, the EMS determines the *Strategy* considering the driving operation and simultaneously sets the *Policy* for optimized use of power sources in HEVs [EWK06, TT12, GS17, HWK$^+$17, ORBE20] or in conventional ICE vehicles [KKJ$^+$07].

The resulting software architecture of the APN management is depicted in Figure 2.2. In contrast to the definition in [RLEW06, RL07] and Figure 2.1, the EMS directly interacts with the PES. Thus, the EMS defines the references values SOC_{ref} of the storage systems and at the same time derives the necessary power and voltage references, P_{ref} and V_{ref}, for the APN [GS17, ORBE20]. As a result, the horizontal layers in the pyramid are broken, and the PMS is situated beside the EMS.

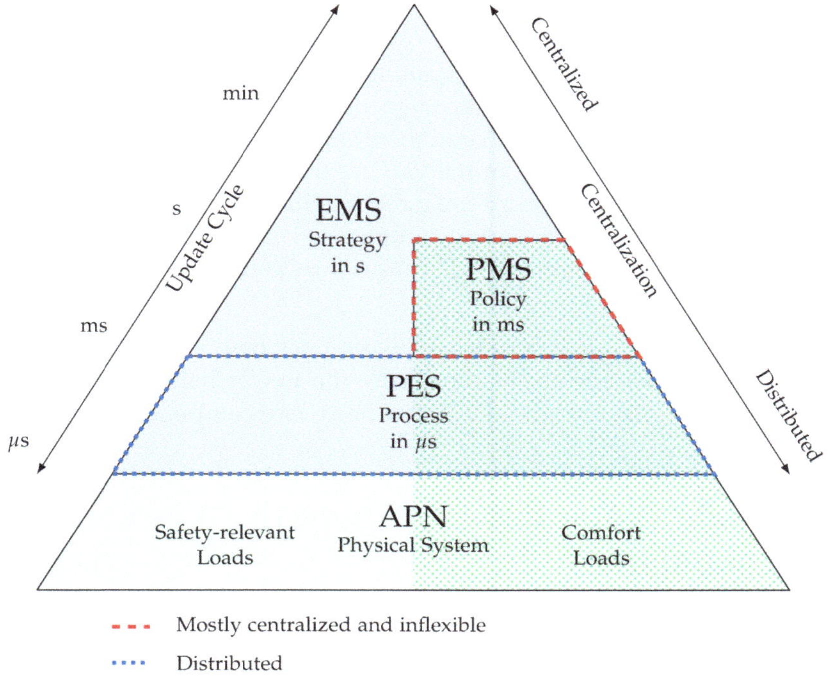

Figure 2.2: New software architecture for the APN management shells with the EMS, the PMS, and the PES. The architecture based on [RLEW06, RL07] is adapted with respect to the elaborated role of automotive power management (PMS) [SGH24].

New Power Management Task

Due to the changing scope of the EMS, which takes over some of the previous PMS tasks, the PMS in Figure 2.2 has new tasks. In this regard, the placement in the management pyramid represents the new PMS functionality as an observer and fallback mechanism as highlighted in [SGH24]. During the general vehicle operation, the EMS determines the long-term strategy. However, the EMS does not take into account the whole APN operation. There are unforeseen events, such as weather changes, or APN states, for example, component temperatures [ORBE20]. Additionally, the human driver and the passengers influence the APN states by determining the driving operation and activating or deactivating comfort components. Furthermore, non-automated driving often involves unpredictable road conditions or environment changes leading to the activation of safety-relevant loads [HWK+17].

If these changes lead to power imbalances, the PMS in its new role interferes with rebalancing the power supply and demand, stabilizing the power supply. In this regard, the PMS ensures the power balance in (2.1) focusing on the comfort loads as highlighted in Figure 2.2 [KET+11, RBW+12]. To fulfill this task, the PMS overwrites the power and voltage references, P_{ref} and V_{ref}, of the EMS.

Regarding the desired flexibility to integrate new hardware components seamlessly, the software architecture of the APN management plays a significant role. For the integration of new comfort loads, the PES and the PMS in its new role are the responsible functions. While the control cycles in the PES are hardware-specific and distributed, the PMS approaches in state of the art are mostly centralized functions that aggregate and process the information in a central computation unit [KKJ$^+$07, Ise16, ORBE20]. Therefore, integrating new comfort loads causes a redesign of the centralized power management.

As defined in the research goal, this work investigates new automotive power management approaches that effectively coordinate the power balance in the APN and facilitate the plug-and-play integration of new hardware components.

2.2 Desirable Properties of Automotive Power Management

According to the proposed automotive power management, the following definitions present the desirable and necessary properties. Furthermore, the definitions are given to clarify the nomenclature used in the thesis. In the development process, the desirable properties build the foundation for requirements regarding automotive power management.

A typical division of algorithms is their hierarchical structure [Sca09]. In this thesis, the terms centralized in Definition 2.1, decentralized in Definition 2.2, and distributed in Definition 2.3 are defined as follows. They represent one aspect of the software architecture within the vehicular E/E architecture.

Definition 2.1: Centralized Algorithm

A centralized algorithm aggregates and processes all necessary information in one central entity.

Definition 2.2: Decentralized Algorithm

A decentralized algorithm includes more than one entity to aggregate and process the necessary information. Additionally, some form of explicit or implicit coordination and information exchange between the entities ensures the overall effectiveness.

Definition 2.3: Distributed Algorithm

A distributed algorithm aggregates and processes all information in the individual entities. Thus, the distributed algorithm is the edge case of the decentralized algorithm. Every component or system has its own computation and communication unit to participate in the distributed algorithm.

To manage a power network, the centralized algorithm in Definition 2.1 needs a direct or indirect connection for communication with every participant in the network. In contrast, the decentralized algorithm in Definition 2.2 needs a direct or indirect connection for communication to every participant in the network via one of the decentralized control entities. Established forms of coordination between the decentralized units include a hierarchical approach, a leader-follower implementation, or coordination through negotiation [Sca09, MLCA11, CSLL13, BBD+14, MN14]. The distributed approach in Definition 2.3 emphasizes the independence of the individual entities and is often referred to as a multi-agent system (MAS) [ISM14, CSAH23].

In the three definitions, the algorithms have decreasing spheres of influence. As a result, the effect of a single computational unit failure ranges from the whole system (centralized) through parts of the system (decentralized) to individual components (distributed). Hence, a respective distributed algorithm has inherent safety precautions [GTL+08, WSD+18].

One major aspect of this work is the power management design enhancing or hindering a plug-and-play integration of new software components, particularly hardware components, into the APN and the management structure. Since centralized algorithms as described in Definition 2.1 are mostly built monolithically, integrating new software components requires high effort and maybe a complete algorithm redesign. Therefore, the modularity of an algorithm or hardware system elaborated in Definition 2.4 provides the possibility for seamless exchange of components [HMS09].

Definition 2.4: Modular Algorithm

A modular algorithm comprises separate parts with clearly defined interfaces. To perform superordinate tasks, the single parts of a modular algorithm exchange information via the interfaces. If the interfaces are kept constant, individual parts are seamlessly exchangeable, adaptable, or extendable without affecting the rest of the algorithm.

The defined modularity is a critical cornerstone for the desired plug-and-play integration of hardware components into the APN.

Definition 2.5: Plug-and-play Integration

The plug-and-play integration property of an algorithm describes the effortless integration of new software components into an existing algorithm. Through a modular design providing clear interfaces and structures, the algorithm integrates new components without the need for adaptations regarding communication structure or algorithmic design.

In this work, the plug-and-play integration property in Definition 2.5 is extended to the combination of hardware and software. Accordingly, if the hardware design and the software design fulfill the property of modularity, the plug-and-play integration property in Definition 2.5 is valid for the integration of new hardware components. Consequently, this affects the power connectors and the communication technology on the hardware side. On the software side, modularity has to be achieved by the communication protocols and the functional integration concerning the APN management system. For the integration of comfort loads into the APN management, this affects, in particular, the PES and the PMS in Figure 2.2.

Besides the APN observation and monitoring, the main power management task is the assurance of a continuous power balance as described in Definition 2.6. The definition is derived from (2.1). For comprehensive and effective power coordination, the power management has to include all possible components $i \in$ APN to achieve the power balance. Additionally, user and passenger preferences play a significant role. As an example, the complete shutdown of all active comfort components in case of a slightly excessive power demand would be regarded as an unsuitable reaction and, in particular, diminish the user comfort.

Definition 2.6: High Quality of Power Balancing

A high quality of power balancing enhances the stable power infrastructure and is determined by a trade-off between load reduction and the increase in generation. Considering passenger preferences and further constraints, load reduction and generation increase have to be applied precisely within the power balancing decisions.

In accordance with the research goal, the plug-and-play integration in Definition 2.5 and the quality of power balancing in Definition 2.6 represent the essential properties of power management. However, there are further advantageous features of capable power management. One property is the scalability in Definition 2.7 indicating the possible APN extension with several components. In contrast, the plug-and-play integration property describes the effort according to the addition of individual hardware components. To facilitate scalability in the APN, the hardware platform has to provide sufficient connectors for the power supply and communication. On the software side, the communication protocols and the power management have to feature limited growth in the communicational and computational effort for every new component.

Definition 2.7: Scalability
Scalability describes the property of an algorithm to add components or tasks without hitting certain limits that cause its disturbance or breakdown. For automotive power management, scalability is particularly critical in terms of electric and electronic components in the power network and number of separated voltage levels.

Definition 2.8: Generalization
The generalization of an algorithm describes its applicability to a broad range of problems [FBL70]. Power management generalization describes the transferability according to components and voltage level architecture.

Due to the numerous vehicle and APN configurations, the generalization in Definition 2.8 is a valuable property of a power management approach. In this context, the broad range of problems refers to different configurations of voltage levels and the presence or absence of various components in the APN.

Definition 2.9: Individualization and Customization
Individualization of automotive power management describes the possibility for individual adaptations regarding the different components and systems in the APN [Büc08, p. 72ff]. Customization enables adaptations toward customer preferences [ASN00].

The individualization aspect in Definition 2.9 corresponds to the generalization property. While generalization allows for the broad application of the power management approach regarding varying components, individualization facilitates the specification to consider individual components within a fixed power management structure. An essential criterion in this aspect is the effort to consider varying system states, for example, the internal battery states, such as state of charge (SOC) and state of health (SOH). Additionally, the customization emphasizes the driver or passenger preferences and thereby enhances the comfort aspect.

Definition 2.10: Fail-Safety

The fail-safety describes the property of resilience regarding single or multiple faults or malfunctions [WSD+18].

Since the supply of electric power becomes a safety-relevant feature of vehicles, the fail-safety property in Definition 2.10 is important for capable power management approaches. An approach is considered fail-safe if it ensures proper functioning in case of single or multiple failures. In this regard, a suitable design may divide the system into separated parts so that, in case of a single failure, the surrounding parts are not or little affected by the respective failure. Therefore, the distributed design and the modularity of the power management are key features for fail-safety. Another opportunity to guarantee fail-safety is the redundant operation of parallel systems, which are responsible for the same functionality [WSD+18, SGLS22].

Definition 2.11: Computational Effort

The computational effort sums up the respective computing operations in all participating entities required to apply the automotive power management procedure [ORBE20].

The increase of the computational effort with every additional component in the APN could be a limiting factor for the scalability of the investigated approaches. Moreover, the computational power in the vehicle is limited by the availability of energy as well as the need for weight and cost reduction [WSD+18]. Although the available computational power is expected to rise, the computational effort described in Definition 2.11, in addition to the previously defined scalability, is a major factor for a successful implementation in a vehicle.

Besides the computational effort, the communicational effort in Definition 2.12 is responsible for the applicability of a power management approach. Comparable to the computational effort, the limitations in weight, cost, and available energy request the reduction of communicational effort. Since the amount of data, as well as the frequency of interaction, determine the communicational effort, both aspects are important as possible factors for the assessment.

Definition 2.12: Communicational Effort

The communicational effort assesses the amount of data required to apply an algorithm. Regarding power management, communication includes information from and commands toward the electric components and the interconnection of entities involved in a decentralized or distributed approach.

Definition 2.13: Cost of Implementation
The cost of implementation sums up the computational and communicational effort as well as other factors, such as development and testing, that increase the implementation effort. All implementation costs may be aggregated in a monetary value, including the hardware costs for computation and communication and the engineering costs for development and testing.

The algorithm's complexity in terms of modeling effort and lines of code is responsible for higher implementation costs as defined in Definition 2.13. Furthermore, the necessity for additional hardware in the vehicle concerning the application causes higher costs that are associated with the respective power management approach. Another factor is the testing effort toward the application.

2.3 Automotive Power Management Approaches

Even though flexibility is relevant on all management levels, this work focuses on the power management layer. Since the PMS mainly controls the comfort components, as visualized in Figure 2.2, it seems to be a reasonable starting point to provide flexibility. As a result, the already distributed PES and the PMS would support the desired plug-and-play integration property of new comfort components, making vehicles adaptable toward customer preferences. Additionally, power management has to feature a high power balancing quality and further properties.

Therefore, the power management approaches in the state of the art are elaborated in detail. The analysis is divided into rule-based, dynamic optimization-based, agent-based, and auction-based and economically-reasoned approaches.

2.3.1 Rule-based Approaches

Rule-based or heuristic approaches provide a simple framework for the implementation of power management schemes [ESG04, TT12, RSSH21]. Since there is no need for a model of the underlying system, rule-based approaches are suitable for complex and nonlinear systems for which an appropriate white-box model is difficult to derive. Instead of the systematic understanding on a physical level, the rule base represents the abstract system behavior translated into comprehensible rules. The rules are mostly derived from expert knowledge of underlying mechanisms and system behavior. More precise rule bases may result from offline optimization procedures adapting the operation ranges or the parameters of the resulting actions. In [WTH18], the energy and power management strategy parameters within a state machine are optimized to suit conflicting targets.

With regard to automotive power management, the rules describe actions referring to certain states of the APN [ESG04]. These actions comprise commands to the generator or the DC/DC-converter to regulate the voltage level. Furthermore, the rules request acute load reductions in cases of insufficient power supply or a low battery SOC [ESG04]. In order to further reduce complexity in the rule base and increase modularity, a rule-based power management approach may divide the system levels, leading to a hierarchical design. Hence, there are rule bases for the overall system and individual rules for the subsystems, such as the single ECUs [Dör16].

The rules generally focus on a stable power supply in the APN. Another application of a rule base is the rise of power network efficiency. In [Dör16], the three aspects ECU degradation, pretended networking, and partial networking are proposed to reduce the power consumption concerning the ECUs. The overall goal is the degradation of computational and communication effort for inactive parts of the APN. The proposed mechanism design includes an observer and a controller for each system level. The observer aggregates the available information and extracts the current state of the system, such as the overall power network or a single ECU. Based on the current state and a predefined knowledge base, the controller derives the next action to apply. Consequently, the controller demands the degradation of systems or particular ECU components to limit power consumption during phases of low communication or computation activity. To show the working principle and effectiveness, the rule-based power management is implemented and validated in an experimental vehicle platform [Dör16]. In [KGT19], the energy saving through ECU degradation is extended by event prediction applying neural networks. The authors consider the breakeven time for ECU degradation in the neural network training, thereby preventing counterproductive ECU state transitions.

An innovative approach with distributed control units and a centralized coordinator is presented in [GYZ+10]. The proposed design features component functions in direct proximity to the hardware and a rule-based power management, which is coordinated from a centralized supervision and coordination unit. However, the architecture mainly focuses on the ECU power supply.

In contrast, the proposed intelligent load management in [KET+11] ensures voltage stabilization by distributed load reduction in the APN. The control algorithm comprises a centralized estimation and coordination that sends control commands for load reduction to the individual components. In [RBW+12], a distributed approach without a centralized coordinator is proposed and validated in simulation. The experimental results in both works demonstrate the effectiveness of voltage stabilization in a 12V power network [RWM+13]. Additionally, in [BFMB12], a software architecture with a centralized power distribution coordinator and respective local subsystems is presented. A similar approach, which introduces a hierarchical structure with local subsystems, is proposed in [KFB+10, Koh14] and facilitates fast communication and predictive measures to enhance voltage stability by local load reduction. Furthermore, a distributed design for the electric and electronic architecture with a central coordinator for power supply is presented. The shutdown mechanism in [RBW+12, Ruf15] allows for distributed measures against acute voltage drops in the APN.

In [KYZ+11], a power management approach for automotive power networks is proposed, dividing the power supply into two separate areas. One area is the supply of the sensitive ECUs, and the other is the supply of the electric loads in the APN, which induce voltage and current changes due to volatile power demands. Consequently, the approach facilitates a reliable supply for the ECUs with little electromagnetic disturbances, over-current protection, and high voltage level stability.

2.3.2 Dynamic Optimization-based Approaches

The application of optimization-based approaches[1] builds upon a thorough system understanding and a mathematical description of the system behavior. Optimization-based approaches provide optimal control performance in terms of the defined objective function [BBM17]. In contrast to the rule-based power management algorithms, which are based on abstract expert knowledge, the design of dynamic optimization-based approaches usually integrates accurate white-box models to describe input-to-output relations or internal system states. With regard to the different shells in Figure 2.2, dynamic optimization-based strategies offer a capable and well-established approach for the EMS focusing on the driving operation [NKK+03]. In a few approaches, the PMS tasks considering auxiliary loads are integrated into the overall problem formulation [KKJ+07, RDKW19].

To determine the optimal input for a given system, dynamic optimization-based strategies solve an optimization problem by calculating the optimal solution, which minimizes a set of objective functions. According to the desired system behavior, the objective function usually includes internal system information, which is weighted to quantify the importance of specific criteria or goals. As a result, optimization-based approaches are more straightforward to scale or adapt than rule-based strategies since the objective function can be adapted to emphasize different goals. However, optimization-based strategies often require high computational efforts to solve the optimization problem and derive the control input [ORBE20]. Besides the objective function, constraints to system inputs or states in the problem formulation provide the opportunity to consider system limitations and restrict the optimal solution to a feasible set [BBM17]. Internal system information, as well as external information about the driver's behavior, the driving cycle, or the weather conditions, are the foundation for an accurate problem formulation and a capable solution [HWK+17, RSSH21].

In [KKJ+07], the authors describe a dynamic optimization-based approach focusing on energy management for vehicles with a conventional drivetrain. The vehicle configuration comprises an ICE to power the mechanical powertrain and propel the ve-

---

1 The terms *dynamic optimization-based* and *optimization-based* are used synonymously in this work.

hicle. Additionally, the ICE drives the alternator, which supplies the power network with electric power and energy. The authors mainly aim to improve the overall vehicle efficiency. In this regard, the proposed problem formulation includes the ICE efficiency for different operation areas, the efficiency of the battery, and the electric loads in the APN. Furthermore, the problem design comprises constraints according to the power and energy consumption of the electric loads as well as battery limitations. The strategy enhances the vehicle efficiency by ICE operating point adaptations enabled by load shifting with regard to electric loads. Since the future driving cycle is usually not predefined, an online model predictive control (MPC) approach is implemented in simulation and a test vehicle, iteratively calculating an optimal solution for the upcoming driving cycle in a preset prediction horizon.

Due to the utilization of load flexibility, the proposed EMS approach in [KKJ+07] comprises elements of the power management tasks located in the PMS in Figure 2.2. Nevertheless, a separate PMS algorithm may be necessary to fulfill the remaining tasks, such as the short-term power balance with an emphasis on electric comfort loads that do not offer flexibility for load shifting and are not regarded in the proposed approach. In summary, optimization-based approaches emphasize energy management [ORBE20, RSSH21] while there are few approaches regarding power management.

2.3.3 Agent-based Approaches

The application of a multi-agent system (MAS) offers advantages regarding flexibility and safety [ISM14, CSAH23]. Since all agents in an MAS comprise the necessary communication and computation capabilities to operate independently, new agents may be added to the system with little effort, resulting in a plug-and-play integration property of the underlying system. Additionally, if an agent fails, another agent may take over the tasks so that the mutual goals can still be reached [GTL+08, ISM14]. In case another agent can not take over the tasks of a failing agent, the impact on the underlying system is limited to the area that the failing agent controls. Hence, a severe single point of failure may be overcome by a suitably designed MAS.

Another aspect of MASs is the reduced computational power of the various nodes in a system. Since the tasks are divided between multiple agents, the individual computational effort can be effectively limited. Important characteristics by which a proposed MAS can be categorized are the communication scheme and the mechanisms to reach the mutual goals.

In [ISM14], an MAS for an electric vehicle's energy and power management is presented. For the proposed MAS, every electric component in the APN is equipped with an individual agent. This agent controls the component actions and negotiates the energy and power consumption with the surrounding systems. Hence, a reasonable power balance in the APN is the mutual goal of all agents in the MAS.

Two mechanisms are implemented to coordinate the power consumption. The first approach is based on a power auction between all agents, coordinated by a central auctioneer agent referring to [Gra04a, Büc08]. The second mechanism applies a decentralized negotiation between neighboring agents, leading to an iterative power trading in the MAS based on the respective component utility. As a result, the power consumption in the overall APN is balanced according to the available power and the power utility for every agent and the overall vehicle operation.

In contrast, the agent-based approach in [REL05] performs a load type segregation which classifies the comfort loads in the APN based on their priority and time constants. During the heuristic algorithm for power distribution management, the available power is distributed between the active loads in accordance with their weighted priority. In [LR05], the same power distribution problem is solved by a negotiation-based approach, which is calculated with a genetic optimization algorithm. The compelling power balancing is demonstrated in a simulative validation with the ADVISOR toolbox[2] [Joh02, MBH+02].

2.3.4 Auction-based and Economically Reasoned Approaches

A promising methodology facilitating the targeted power management flexibility is the auction-based approach first presented in [Gra04a] and patented in [Gra04b]. In energy markets, auction-based and economically reasoned approaches have a broad range of applications and are still a vivid area of research [MBC11, MDSH14, TCY+16, NZ20, DCK21, Mau23]. The basic idea is to treat the electric power in the APN as an economic good comparable to the mechanisms in energy markets [MBC11]. On the one hand, power suppliers provide power capabilities and offer electric power as an economic good. On the other hand, some consumers demand electric power to provide certain functions, such as air conditioning, heating, or window lifting. In the auction mechanism, the two categories act as power sellers and power bidders. Consequently, auctions depict possible mechanisms or protocols to determine the price of the traded electric power and thereby set its distribution among the bidders. For the fundamentals, theoretical insights, and practical examples regarding auction theory, it is referred to [Kle04, LM05, MBC11, NZ20].

Compared to the application in the energy market, the author in [Gra04a] adapts the UPA for the APN. Thus, the auctioneer in the specified auction determines a uniform price for the electric power and thereby clears the market and achieves a power balance [NZ20]. For the UPA design in [Gra04b] and [Büc08], every seller and bidder has an individual price-to-power function (PPF), which defines the relation between the market price p and the provided or demanded power P. To clear the

2 The ADVISOR toolbox is written in MATLAB/Simulink and has been created by the National Renewable Energy Laboratory as a comprehensive vehicle simulation which is used to simulatively assess different powertrain configurations including batteries, ultracapacitors, fuel cells, and conventional ICEs [MBH+02].

market and determine the current price p, the auctioneer calculates the intersection between the power supply, growing with an increasing market price, and the power demand, decreasing for an increasing price [Büc08].

In [Büc08], the auction-based approach from [Gra04a] is adapted and explained in detail regarding the mathematical and economic background of auction theory. Additionally, the author presents further mechanisms based on economic concepts that work similarly to the auction-based approach. Different PPF properties and their influence on the market price adaptation and specific auction designs are discussed for the auction-based approach. Subsequently, the proposed mechanisms are integrated into an abstract power and energy management architecture comprising various software components for the central auctioneer and the distributed sellers and bidders of electric power in the APN.

2.4 Neighboring Methodologies and Domains

The most obvious neighboring domain is the energy market, in which comparable challenges regarding energy and power supply occur. The general design and different layers in the energy market are similar to the automotive management structure shown in Figure 2.1. Therefore, the technical solutions in the energy market provide suitable approaches for the APN. Due to only slight differences between microgrids and the power network in today's vehicles, some concepts from the energy market are transferable with little adaptations.

One example is the demand side management described in [PD11] and [MRWJ$^+$10], which mostly translates into rule-based approaches. However, the implementation of rule-based mechanisms in the vehicle is usually centralized, lacking the desired flexibility in the power network. Further concepts, such as pricing mechanisms, load shifting, load shaping, or load reduction, build the theoretical foundation for the previously presented algorithms and mechanisms in the auction-based, agent-based, or economically-based approaches [Büc08, PD11].

With regard to other transportation systems, there are few power management approaches within the state of the art. Due to the high fail-safety requirements, the power infrastructure in airplanes comprises significant differences compared to the APN. In contrast to the APN, the airplane power infrastructure contains primarily point-to-point supply connections [TPLS12]. However, there is a need for new distribution concepts since the electric load on board increases. For electrified ships, state-of-the-art power networks are similar to electric microgrids [FBPZC11, FBPZ15, CPS$^+$15]. In [FBPZ15], the authors propose an agent-based approach for electric load management in the different power network domains. The implemented MAS with its distributed agents controls different loads and generators in the four zones of the electric microgrid comprising AC (alternating current) and DC (direct current) power levels. Further power balancing approaches for all-electric ships are presented in [CPS+15]. For trucks or buses, most state-of-the-art approaches primarily focus on energy management for hybrid propulsion concepts [DLCL21]. In contrast, the authors in [RDKW19] propose an optimization-based energy management approach for trucks, which includes particular auxiliary loads in the optimization procedure. In [CKW15], an adaptive approach utilizing game-theoretic strategies is presented for heavy-duty trucks. The energy management approach supports adaptation to different driving patterns, while the game-theoretic strategies include electric auxiliaries.

The analysis in [WWWMM01] highlights different auction protocols to solve distributed scheduling problems. In this regard, the bidders in the auctions are modeled as autonomous agents. Thus, there is a strong connection to the previously described agent-based approaches [LM05].

Another possible methodology focusing on distributed decision-making is the bargaining approach in [CSAH23]. The authors present a MAS design for active battery balancing, similar to the agent-based approaches. In the proposed battery system design, every cell or module has the capacity for communication with the neighboring cells and the computational power to make individual decisions. Since there are no predefined balancing procedures, the cells or modules negotiate with their direct neighbors to form coalitions in which electric charge is exchanged for the next period. In this regard, multiple optimization problems are derived from the bargaining process, which are solved in every distributed computational unit. As a result, the battery system as a whole provides the plug-and-play integration property for integrating new cells according to the active balancing algorithm.

Methodologies that belong to the category of dynamic optimization-based approaches are the decentralized and distributed MPC algorithms in [Sca09]. In general, these methodologies are advantageous to control large-scale and complex systems that are divided into separate subsystems. These subsystems comprise individual states and actuators. However, the coupling between the subsystems has to be considered for an adequate control result [Sca09].

In [MLCA11], a negotiation-based approach for the coordination of decentralized MPCs is presented. Further, the local cost optimization in [LZ16] provides a decentralized MPC implementation with a low communication load. In the local cost optimization, every decentralized controller optimizes the objective function for its local subsystem. Thus, the communication effort is relieved, but, on the other hand, the global performance of the overall system is limited [LZ16]. As an improvement according to the global performance, the authors in [LZ16] proposed networked MPC approaches that take into account the objective functions of neighboring subsystems or the global objective function.

The hierarchical structures in [CSLL13] build a direct communication approach for the required coordination and synchronization. Another alternative is the decomposition in [BBD+14], which results in an individual MPC approach for each subsystem and a global optimization agent that accounts for global constraints. A drawback, according to the application of decentralized and distributed MPC approaches, is the APN structure, which comprises only the voltage levels as clearly divided subsystems. Hence, the degree of flexibility and modularity remains very coarse, thereby not supporting the effortless integration of individual components. If the tendency toward more PEs in the vehicle rises, these approaches may be more suitable for the APN management [EWK06].

Another neighboring area is elaborated in [BBM00] and deals with the power management approaches in large computational networks or electronic units. The work presents different methods to manage and balance power consumption in such large networks. A key factor is the trade-off between the power and performance cost for state transitions, on the one hand, and the savings for lower system states, on the other hand. Hence, the goal is to switch off components in idle mode but simultaneously consider the possible need for computational power in the following instances. Another aspect in [BBM00] is the suitable design of future electronic systems to facilitate effective power management. For the APN, there are research works on similar procedures to limit the power consumption of communication systems that are not active [Dör16, WSD+18]. These degradation mechanisms and other approaches to reduce energy consumption in the vehicle are further surveyed and assessed in [SKSS10].

2.5 Qualitative Comparison

The following comparison comprises nine categories to assess the different automotive power management approaches. The categories are defined in Section 2.2 and include the desirable properties of capable power management. Table 2.1 provides an overview with respect to these categories and thereby summarizes the assessments with an emphasis on the four algorithm classes: rule-based, dynamic optimization-based, agent-based, and auction-based. The assessment results build the foundation for the research gap, which is elaborated subsequently.

1) Quality of Power Balancing

The quality of power balancing, as described in Definition 2.6, is the main task of automotive power management. In this category, all approaches achieve high effectiveness. Nevertheless, optimization-based, agent-based, and auction-based management perform a more precise power balancing than the simple rule-based approaches and support a more holistic procedure. Since the proposed rule-based approaches in [ESG04] and the shutdown mechanism in [RBW+12] allow for a distributed design, these measures are a promising concept to implement a highly responsive countermeasure for power imbalances and voltage drops. However, the distributed design

Table 2.1: Qualitative assessment of the elaborated categories of power management approaches with the range from advantageous (+ +) through neutral (o) to disadvantageous (- -).

	Approach	Rule-based	Optimization-based	Agent-based	Auction-based
1)	Quality of Power Balancing	o	+ +	+	+
2)	Plug-and-Play Integration	+	-	+ +	+ +
3)	Scalability	+	- -	-	+ +
4)	Generalization	+ +	-	+	-
5)	Individualization and Customization	o	+	+ +	+
6)	Fail-Safety	+ +	-	+	- -
7)	Computational Effort	+ +	- -	-	+
8)	Communicational Effort	+ +	o	- -	- -
9)	Costs of Implementation	+	-	- -	o

and the locally limited effect in the power network impair a holistic power management approach, resulting in a diminished quality of power balancing. In the auction mechanism, the reasonable design of the protocols for selling and bidding plays an important role. For MAS approaches, the actual implementation of the power management algorithm is not defined. Thus, the quality depends on the negotiation or coordination mechanism and varies respectively [REL05, RLEW06, ISM14].

2) Plug-and-Play Integration

This work focuses on the plug-and-play integration as defined in Definition 2.5. The plug-and-play integration reduces the development and design efforts for new vehicle configurations and variants. Additionally, it allows for seamlessly integrating hardware components after production. In the following, the power management approaches are assessed regarding their fulfillment of the plug-and-play integration property.

The optimization-based approaches hinder an easy integration of new hardware components through their monolithic algorithm design and the objective function, commonly adapted to the specific APN [KKJ+07]. In the rule-based approaches, new components can be considered in new rules. However, the entire underlying rule base may need adaptions for significant changes, leading to additional effort during development and with every hardware change after production. In contrast, the agent-based and the auction-based approaches support plug-and-play integration. While in the agent-based approach, a new agent has to be designed and integrated into the MAS [Ise16], the auction mechanism provides the greatest flexibility toward integrating a new hardware component. Concerning the auction, a new component is simply added as a further auction participant [Büc08]. However, the integration of new participants and the referring modularity have to be ensured in the central clearing unit of the UPA. In contrast, the integration of new agents in the agent-based approach is facilitated by the underlying communication framework between all agents in the MAS [Ise16].

3) Scalability

For optimization algorithms, the calculation effort commonly increases with the number of variables [BBM17]. Hence, the optimization-based approaches impair the power management scalability. Therefore, in [KKJ$^+$07], the authors apply an MPC algorithm to achieve a real-time capable design. In a rule-based setting, every new component may result in an additional rule but causes little computation effort. In particular, the auction mechanism in [Gra04a] and [Büc08] facilitate high scalability since new components are easily added to the auction participants. Afterward, the electric power is distributed according to the new constellation of participants in the auction. For agent-based power management, the scalability depends on the underlying coordination mechanism. If this mechanism applies some form of auction, the scalability would be respectively high. However, if the agents negotiate with each other, the computational and communicational effort may rise with every new component [ISM14, CSAH23].

4) Generalization

Since an optimization problem is commonly specific for an individual APN setup and certain components, the generalization of optimization-based power management approaches is limited [KKJ$^+$07]. However, for similar APN configurations, the necessary adaptations for a high power balancing quality may be acceptable. Thus, there may be a trade-off between the quality of power balancing and the demand for generalization.

In contrast to the optimization problems, rule bases are easier transferable toward a slightly changed APN since most rules concern individual loads, which reduce their consumption in certain situations. This procedure is rather independent of the underlying hardware configuration [ESG04]. In the auction-based approach, the algorithm has to consider the power availability in the APN, which may vary between the different voltage levels [Büc08]. For an MAS, the coordination design with direct interaction between neighboring loads provides a high generalization. However, if the agent-based approach includes a central optimization or coordination mechanism for the power distribution, the generalization is comparable to the optimization algorithms [ISM14].

5) Individualization and Customization

Since the rule-based power management design applies a predefined knowledge base and commonly fixed actions, the adaptation to individual component characteristics or user preferences results in great effort. In contrast, the optimization-based approach has the possibility to adapt the weighting of the objective function to adjust its behavior accordingly [KKJ+07, BBM17]. Due to the given negotiation or auction framework in agent-based and auction-based power management, individual agent or participant behavior is changeable without compromising the overall mechanism. Thus, these methods allow for flexible component behavior while the defined framework ensures the power balancing [Gra04a, ISM14].

6) Fail-Safety

With the centralized auctioneer, the auction mechanism has a single point of failure that impairs the fail-safety property [Büc08]. Additionally, failures and errors of individual auction participants may disturb the proper auction mechanism. In [KKJ+07], the optimization is performed in a centralized manner. However, some methodologies apply a more decentralized or even distributed optimization [MLCA11, MN14]. In the rule-based and agent-based approaches, a distributed design is possible and depends on the implementation. In this regard, [RBW+12] proposes an autonomous load shutdown directly implemented in the distributed ECUs. According to the coordination among the MAS, agents may negotiate with their direct neighbors to achieve a power balance in the APN [ISM14, CSAH23] resulting in a single point of failure prevention. In case of a single agent failure, the remaining network is still functional as long as the communication between the remaining agents is not disturbed [GTL+08, WSD+18].

7) Computational Effort

While the rule-based and the auction-based mechanisms need little computational effort, the optimization-based and the agent-based approaches require a certain computational power to provide real-time capable performance. In optimization-based approaches, the application of MPC facilitates the limitation of computational effort by a tailored receding horizon and, thus, a real-time capable implementation

[KKJ+07, BBM17]. The actual necessary computation power for agent-based coordination strongly depends on the applied coordination and negotiation method. In this aspect, the auction-based or heuristic coordination represents a low-effort approach. At the same time, central optimization with a genetic algorithm can lead to high computational needs [RLEW06, CSAH23].

A side aspect of the computational effort is the division between the different computational units. For example, the necessary computational power for a decentralized or distributed approach may be homogeneously spread in the agent network [ISM14, CSAH23]. However, most distributed approaches require, in total, more computational power since every unit has to provide the basic functionalities, such as communication abilities, for proper operation.

8) Communicational Effort

In terms of communication, the rule-based approaches require the least effort. In particular, the autonomous load shutdown in [RBW+12], which performs a local load reduction without further coordination, omits communication. In contrast, the coordination and negotiation between agents or auction participants generate a high frequency of communication within the APN [Gra04a, Büc08]. To adequately control the components in the APN, optimization-based approaches cause an iterative communication to the underlying systems in both directions [KKJ+07]. However, for a centralized optimization algorithm, there is no need for coordination between different entities.

9) Costs of Implementation

Among the presented approaches, salient differences are partly considered in the computational and communicational effort aspects. Nevertheless, the costs of implementation represent a valuable category with regard to the applicability.

The most straightforward power management is the rule-based approach, which comprises a knowledge base with actions performed accordingly [ESG04]. If the implementation is locally constrained, the effort and costs may even decrease [KET+11, RBW+12]. However, the testing process for the rule-based approaches requires high efforts since the effectiveness of the rule base for all driving situations has to be guaranteed [RWM+13]. Therefore, the testing efforts increase the implementation costs. Contrarily, the other approaches provide a solid framework that guarantees a particular system behavior and, thus, reduces testing efforts in the long term.

On the other hand, the optimization-based, agent-based, and auction-based designs comprise extensive coordination mechanisms with partly increased computational effort. As a result, the optimization-based approach requires additional computation power and a sophisticated implementation in order to facilitate a real-time capable application [KKJ+07, BBM17, HWK+17]. For a broad MAS in the vehicle, there is the need for suitable ECUs in proximity to the various electric components in the APN. These ECUs need the computational and communicational capabilities to participate in the respective coordination mechanism representing the electric components as an agent [GTL+08, ISM14]. The auction approach in [Gra04a] and [Büc08] supports a simple implementation but requires communication capabilities concerning all participants in the auction.

2.6 Research Gap and Contributions of the Thesis

In conclusion, the state of the art comprises a broad range of power management approaches with different characteristics. While rule-based [ESG04] and dynamic optimization-based [KKJ+07, RDKW19] approaches dominate the application in series vehicles, there are several research works toward alternative methodologies, such as agent-based [LR05, ISM14, Ise16] or auction-based mechanisms [Gra04a, Büc08], providing inherent flexibility. This flexibility is a crucial basis of the desired plug-and-play integration property. Furthermore, several methodologies from neighboring domains offer suitable measures for automotive power management.

In the literature, most research works focus on energy management and, in particular, on HEVs with different energy sources [TT12, GS17, ORBE20]. In comparison, automotive power management ensuring the power balance in the APN is a seldom research object. This may be reasoned mainly by the higher potential for energy management optimization in terms of energy efficiency and driving performance [KKJ+07, RDKW19, ORBE20].

Regarding Table 2.1, all presented approaches have certain advantages and disadvantages concerning the various power management tasks and properties. The emphasized plug-and-play integration of new hardware components and the related scalability are mainly considered in the agent-based and auction-based mechanism design. For example, in the auction procedure, an arbitrary amount of additional components can be included effortlessly as auction participants, which supports plug-and-play integration as well as scalability [Büc08]. On the other side of the spectrum, the dynamic optimization-based algorithm in [KKJ+07] has a monolithic design, resulting in high efforts for the consideration of additional APN components. For agent-based power management, the scalability strongly depends on the applied algorithm for power balancing. If the agents freely negotiate with each other, the computational and communicational effort increases exponentially with every agent in the network [ISM14, CSAH23]. Thus, every hardware component with its agent puts an additional burden on the power management system. Furthermore, implementing an MAS is a complex endeavor since every agent needs a basic setup of skills to participate and negotiate within the network [Ise16]. In contrast, the auction-based mechanism in [Gra04a, Büc08] offers a simplistic approach with a solid negotiation framework.

Therefore, auction-based power management has been chosen for further investigation. It features high-quality power balancing, effortless plug-and-play integration, and great scalability.

The auction-based approach was first presented in [Gra04a] and [Gra04b]. As elaborated in [Büc08], the approach offers the possibilities for extensions and adaptations in terms of mechanism design and individual PPFs. However, previous works do not provide a consequent mathematical formulation of the auction-based power management approach. Moreover, the proposed procedures have not yet been extensively validated either in simulation or on a test bench. Since the auction-based concept in [Gra04a] and [Büc08] only considers a single voltage level in the APN, it is not applicable to current power network architectures with multiple voltage levels and interconnections with DC/DC-converters. Hence, the generalization suffers due to the limitation to a single voltage level. Another major factor is the communication design, which has been neglected in the previous conceptual ideas given in [Gra04a, Gra04b, Büc08]. In general, the automotive communication schemes play a critical role in the properties listed in Table 2.1 and have usually been regarded as separate from the APN management system. In conclusion, the development of the auction-based concept comprises the following steps:

- ◇ The mathematical formulation of the UPA for APNs,
- ◇ the generalization concerning multiple voltage levels within an APN,
- ◇ and a suitable communication design.

As depicted in [KET+11] and [Koh14], predictive measures facilitate effective power balancing and the reduction of severe voltage drops. Therefore, this work aims for a predictive extension that enhances the power balancing capabilities of the auction-based approach. In this context, the shifting of power demands, as presented in [PD11] for the energy grid and in [ÅEB+04, VA15] for auxiliary systems and the HVAC system, has a great potential.

Contributions of the Thesis

In summary, the auction-based concept in [Gra04a] is a promising idea to provide a flexible and holistic power management design facilitating the desirable plug-and-play integration and high scalability. Nevertheless, this concept has to be developed and evaluated for application in modern and future APNs with multiple voltage levels. With respect to the research goal defined in Section 1.1, Figure 2.3 highlights the contributions of the thesis.

First, a *new software architecture of the APN management* shells with a thorough task description for the automotive power management is proposed. Subsequently, the important terminologies and the desirable properties of automotive power management are defined. Based on a *qualitative comparison* of the state-of-the-art approaches, the auction-based concept, first presented in [Gra04a], is chosen for further investigations and developments.

The consequent mathematical formulation of the *auction-based power management* extends the previous works in [Gra04a, Gra04b] and [Büc08]. Furthermore, the approach is improved to suit APN with multiple voltage levels, connected via DC/DC-converters. Introducing individual price-to-power functions (PPFs), the proposed power management supports individual component behavior and the consideration of user preferences. On the other hand, the defined design rules for PPFs ensure a robust auction framework for reliable power balancing. To facilitate the seamless plug-and-play integration of new hardware components into the auction mechanism, a novel *communication design* based on the SOA is developed. For these developments, the fundamentals about different voltage level architectures, individual components and systems in the APN, and available communication technologies build a solid foundation.

Furthermore, a *predictive extension* to the proposed method is introduced, aiming for enhanced power balancing capability. The basic idea is similar to [KKJ+07], [KFB+10], or [VA15]. In [VA15], the authors design an MPC to deploy the HVAC system as a counterpart to the electric machine and thereby guarantee a reduction in peak power. The predictive extension complements the proposed auction-based power management. It makes use of the HVAC system's flexibility to shift or shape future load demands, achieving a smooth power balance in the APN.

Finally, a first simulative *evaluation* points out the effectiveness of the auction principle in the APN. The simulations comprise relevant driving situations with highly volatile power demands that call for effective power balancing. However, the simulations neglect communication between the electric and electronic components in the APN, focusing on the fundamental auction procedure. Consequently, the simulation results demonstrate the working principle of auction-based automotive power management and the predictive extension. Thus, the simulative evaluation in this work creates the foundation for further developments and validation processes toward the application in test benches or experimental vehicles.

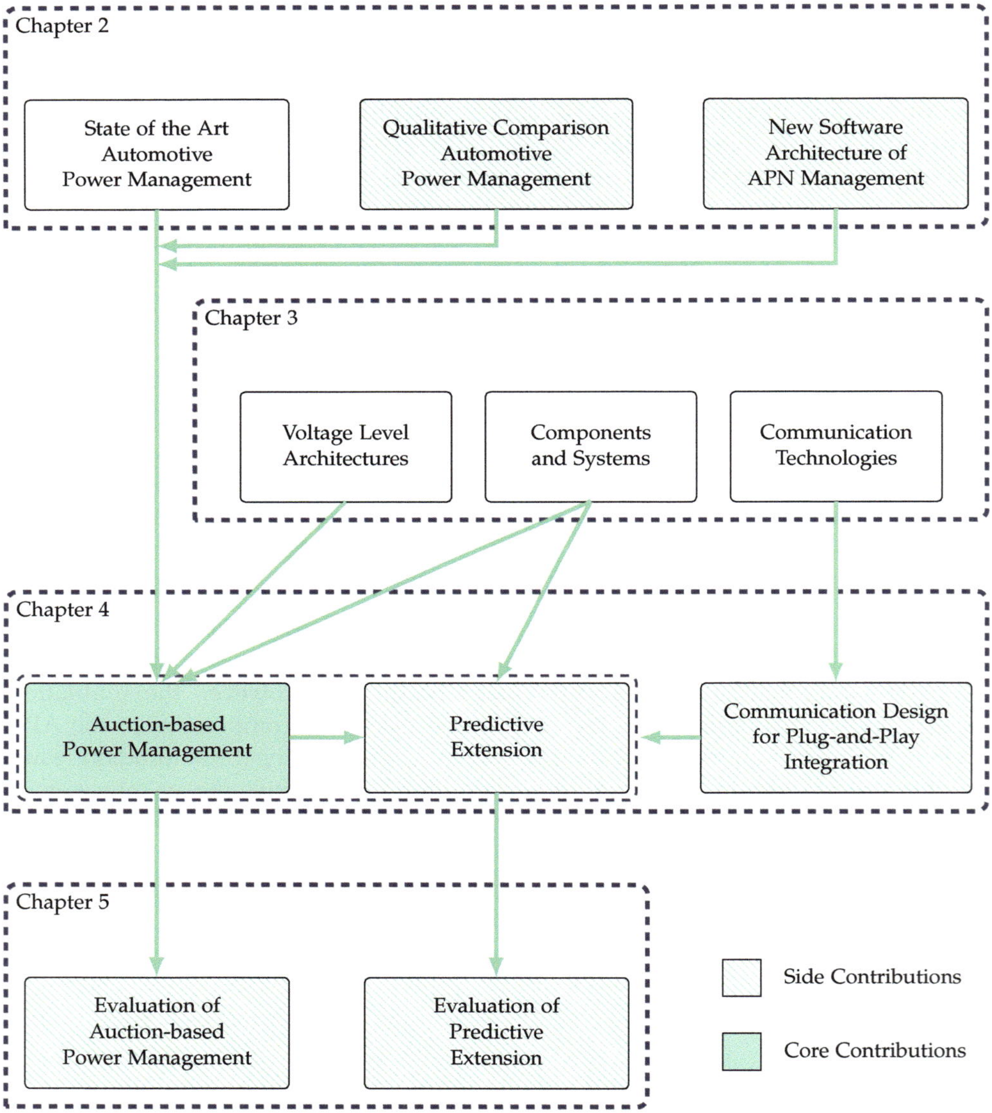

Figure 2.3: Structure of the thesis comprising the research subjects in the chapters, the side, and the main contributions highlighted by the green patterns.

3 Background and Related Work in Automotive Power Networks

The APN, as defined in this work and visualized in Figure 1.1, comprises all physical electric and electronic components in the vehicle. Due to the electrification of automotive components, the rapid dissemination of further telecommunication and entertainment systems, and autonomous driving functions, the number of electric and electronic components has strongly risen in recent years and is still growing [WSD+18, MPPRM+20, TVE+22]. As a result, the interplay of systems in the APN as well as the requirements regarding power and energy supply increase [KWT+10].

This chapter gives an overview of recent and modern APN architectures in Section 3.1, focusing on the different voltage levels and the deployed electric storage systems and energy sources. Subsequently, the important components in the APN, their functionality, and their physical behavior are introduced in Section 3.2 as fundamentals for the later power management design. Section 3.3 presents the current communication technologies and their characteristics, building the foundation for the power management communication design. The chapter concludes with goals toward a capable APN concerning today's and future vehicles in Section 3.4 and a summary of the key findings in Section 3.5.

3.1 Voltage Level Architectures

With regard to the research goal, the APN architecture has a strong influence on the electric power supply in the vehicle. For example, the location of the energy sources or storage systems within the APN harness, as well as the transfer structure for electric power, are mainly responsible for short-term shortages in power supply resulting in severe voltage drops [GFKH09, RWM+13]. Hence, every architecture has strengths and weaknesses according to power transfer, which are determined by design [Hoh10, Ruf15]. From the generalization perspective in Section 2.5, automotive power management has to adapt to these different voltage level architectures and their respective properties.

Single 12V Level

Historically, the ICE propelled vehicle has a power network with a single voltage level of 12V as depicted in Figure 3.1 [SA03, ESG04]. This power level supplies all electric

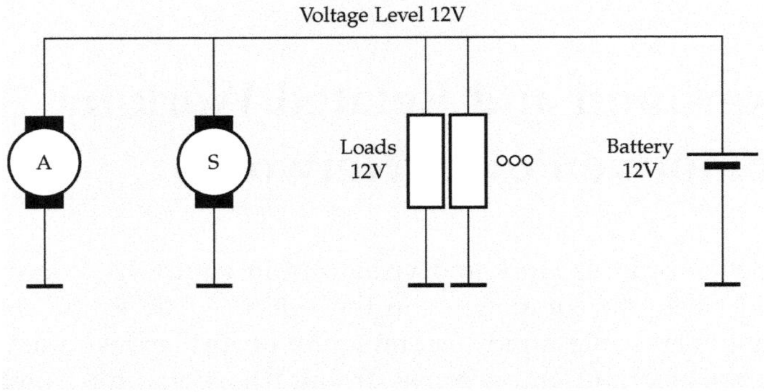

Figure 3.1: APN of a conventional vehicle with one voltage level and the alternator (A) propelled by the ICE, the starter motor (S), and the 12V battery for power and energy buffering.

and electronic components. The lead acid battery (Battery 12V) and the alternator (A) provide electric power for all components within the power level. The alternator is the primary source of electric power. It is mechanically coupled to the ICE and transforms mechanical power into electric power. At the vehicle start, the lead acid battery supplies the starter motor (S) with electric power to start the ICE. This initial start of the commonly cold engine puts high demands on the battery. For the mutual operation during the drive, the alternator provides the electric power while the battery acts as a buffer for a smooth power supply [Büc08].

Two Voltage Levels with 12V and 48V

With the upcoming electrification of former mechanically powered components, such as the HVAC system, and the introduction of HEVs, the power and energy requirements exceed the capabilities of the traditional 12V power network. In order to reduce the necessary currents and increase the efficiency, the 42V or 48V power level was added to the APN [SA03, EWK06, ELR08, BAS19]. Hence, the according APN comprises a power network with two voltage levels connected by a DC/DC-converter. Figure 3.2 shows the physical APN from Figure 1.1 as a common APN architecture with a 12V and a 48V power level for an HEV. A lead acid battery (Battery 12V) still supplies and stabilizes the 12V level, while more capable battery technologies were introduced for the 48V level as the primary electric energy storage (Battery 48V) [ORBE20]. In most cases, the DC/DC-converter supplies the 12V level and takes over the former role of the alternator. Furthermore, an electric machine (EM) in the 48V level provides electric power or additional mechanical torque in generator or boosting[3] mode, respectively. Another feature enabled by the EM capability is the possi-

[3] In boosting mode, the EM produces torque to support driving operation. With a high degree of hybridization, the EM is able to solely propel the vehicle while the ICE is switched off [EGLE18, p. 113ff].

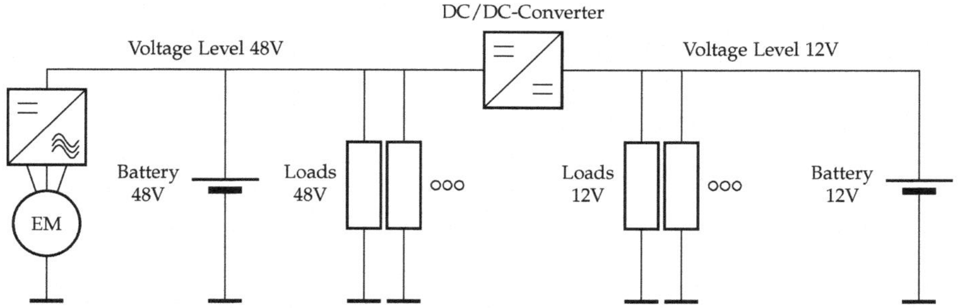

Figure 3.2: APN of an HEV with two connected voltage levels and the electric machine (EM), the DC/DC-converter, and the batteries in both voltage levels as the energy and power infrastructure.

bility of regenerative braking. By the application of negative mechanical torque, the EM decelerates the vehicle and simultaneously converts the vehicle's kinetic energy into electric energy [ELR08], [EGLE18, p. 377ff]. As a consequence, the EM commonly replaces the starter motor, provides the starting moment for the ICE, and creates the possibilities for efficient driving performance.

Multiple and Higher Voltage Levels

Nowadays, for EVs or HEVs with high electric propulsion power, the APN includes even higher voltage levels ranging from 200V to up to 800V [BFB+15, ABK+21]. With the new power levels, current APNs differentiate into several architectures of power networks with up to three or more separate voltage levels [Büc08, BFMB12, GKO19]. The different architectures describe the combination and connection of power levels via DC/DC-converters as well as the placement of electric storage systems and energy sources [ESG04]. For the constellations of voltage levels, combinations exist with 12V, 48V, and 200V to 800V [BFB+15, ABK+21]. In Figure 3.3, an APN with two voltage levels for a standard EV configuration is depicted. The necessity for the different voltage levels results from the standardized component design, on the one hand, and from the power demands of the various components in the vehicle, on the other hand [Rei11, Rei14]. Thus, the complete change from 12V to 48V takes time and causes costs as well as development efforts for the automotive industry.

As energy storages, the APN for EVs and HEVs includes batteries in the different voltage levels or ultracapacitors connected by a DC/DC-converter enhancing the short-term power supply. Additionally, EVs may comprise a fuel cell as a locally emission-free energy source to supply the electric powertrain [EGLE18, p. 421ff]. This would add further complexity to the APN. In this research work, a fuel cell energy source

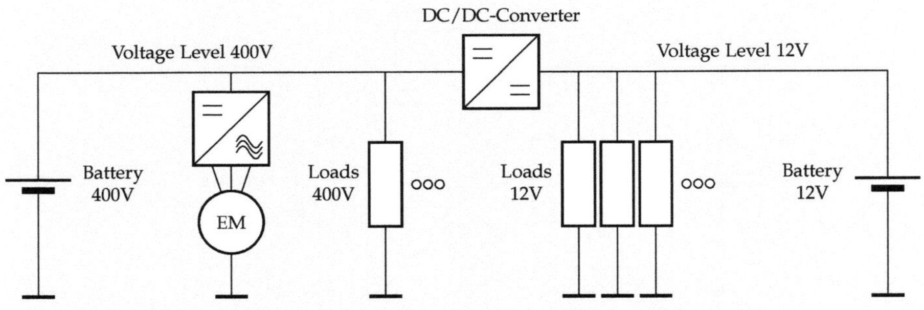

Figure 3.3: APN of an EV with two connected voltage levels and the electric machine (EM), the DC/DC-converter, and the batteries in both voltage levels as the energy and power infrastructure.

is not regarded since it poses additional requirements but does not alter the general power management procedure.

Redundant Voltage Levels and Alternative Designs

Due to the safety requirements for autonomous driving operation, new architectures with redundant power levels are a viable solution to guarantee the power supply in all situations [ABJG12, BFB⁺15, Kir19]. Since sensors, computational units, and actuators, such as the electric steering motor, rely on stable electric power for adequate functioning, these safety-relevant components are partly supplied by two separate power networks [Kir19]. The redundant power networks are kept charged and prepared during the vehicle lifetime but may never be deployed unless the primary power supply fails [SKG17, GKO19].

In [TTFH18] and [TPFH21], the redundant power levels are replaced by a ring structure and further hardware measures, such as decoupling through DC/DC-converter, to ensure the power supply of safety-relevant systems. Through the ring structure, the proposed power network design prevents single points of failure concerning the critical components for vehicle safety. As a result, there are no backup storages or parallel power network structures as a redundant fallback solution. A similar idea with multiple DC/DC-converters connecting two power levels is demonstrated in [WFF⁺16].

In [KYZ⁺11] and [YKLL16], the authors propose an experimental power supply architecture that separates the loads from their control units (ECUs) in order to achieve a stable voltage level for the control units free from the disturbance caused by peak power demands of the loads. Comparable ideas, among other hardware measures, are proposed in [Hoh10] and [Ruf15].

3.2 Components and Systems

The APN comprises different components and systems [Rei14, Nea20]. In this work, some components are categorized as the power infrastructure systems, representing the backbone of the electric power and energy supply. This group comprises the electric machine (EM), the DC/DC-converter, and the electric storage systems. Additionally, the loads are differentiated into loads with safety-relevant or comfort functionalities for the vehicle.

Regarding the power supply and demand, the relevant properties of the components are the average power consumption or supply, the peak power, and the transient behavior. Since the transient behavior constitutes fundamental requirements for the power supply, this dynamic behavior concerning voltage and current and the respective time constants is of particular importance. This is true for the demand side components as well as for the supply side. Since this thesis focuses on the active components of the power infrastructure, the harness, the electric fuses, and other passive components are not regarded and described in detail. For further information, it is referred to [Ema05, Rei14, Nea20].

3.2.1 Electric Machine and Inverter

The EM or the alternator and starter motor are the central components of the electric energy and power network. They represent the primary connection between the electrical and mechanical parts of the vehicle. Historically, the interaction with the mechanical part was separated into the starter motor for the ICE start and the alternator, which consistently converts mechanical torque from the drivetrain into electric energy [ESG04]. In modern vehicles, the capable EM carries out the coupling in both directions. In an HEV configuration, the EM mostly provides electric energy for the APN in generator mode while it occasionally supports the driving operation with additional torque in boosting mode. The relation between electric power generation and boosting is often referred to as the degree of hybridization [EGLE18]. Another advantageous feature of the HEV configuration is the possibility for regenerative braking. In this recuperation mode, the EM decelerates the vehicle and converts mechanical into electrical power.

For starter motors and alternators, mostly direct current (DC) machines are deployed and located in the traditional 12V level of conventional vehicles. Since these machines have a long history of advancements and improvements, present starter motors and alternators are highly efficient and adapted for their respective application area in the vehicle [Büc08]. Nevertheless, deploying alternating current (AC) machines brings improvements for vehicles. In comparison with two DC motors in the single voltage level configuration in Figure 3.1, one AC machine combined with a DC/AC-inverter fulfills both functions, the electric power generation and the ICE start. In HEVs, the

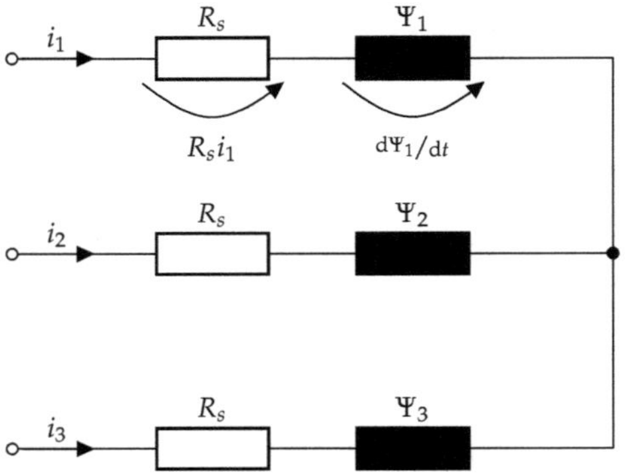

Figure 3.4: Electric circuit diagram of the electric machine stator with the general stator resistor R_s and the magnetic flux Ψ_i in each phase.

EM additionally supports the driving operation with the boosting mode or facilitates a higher driving efficiency through regenerative braking. As a further advantage against the DC motor, the AC machines offer higher efficiency, higher speed, and a more compact volume and size [KWOH17]. Furthermore, the advanced control capabilities of the inverter provide accurate controllability regarding voltage, current, electrical torque, or speed. Inverters support high precision to determine the machine's voltage and current [KWOH17, EGLE18].

According to the AC machine design, the synchronous and the asynchronous motors are the mainly distinguished machine concepts. In general, both motor types comprise a rotor and a stator, with the fixed stator and the rotor usually rotating within the stator. For both motor types, the coupling of magnetic fields in the stator and rotor creates the electrical torque. The electric behavior of the machine is described by the respective relation between current and voltage in the three stator phases depicted in Figure 3.4.

While R_s represents the stator winding resistance, the magnetic flux Ψ_i is induced by the magnetic coupling between stator and rotor, the rotor movement, and the magnetic coupling between the three phases $i \in \{1, 2, 3\}$ [KWSK02, Gem15]. Due to the relation between current and voltage, there is an electric time constant

$$\tau_{\text{EM}} = \frac{L}{R_s} \tag{3.1}$$

caused by the magnetic flux Ψ_i (L), which results in a delayed response in terms of volatile power demands [Hoh10, Gem15, KWOH17]. The time constant τ_{EM} displays the time delay of the electric current i_i compared to the voltage u_i.

The differentiation between the two main machine types concerns the creation of magnetic fields and, thereby, the construction of the rotor. In the asynchronous machine, the alternating current in the stator creates a certain magnetic field. This stator field then induces an alternating current in the rotor, producing the magnetic field. Since electromagnetic induction is caused by the difference between the stator field frequency and the rotating frequency of the mechanical rotor, this motor concept requires a permanent frequency gap to produce electric torque. This difference in speed, called slip s_{EM}, is expressed by

$$s_{EM} = \frac{n_s - n_r}{n_s}$$

and comprises the stator field speed n_s and the mechanical rotor speed n_r in rounds per minute (rpm).

In contrast to asynchronous motors, a synchronous motor does not rely on the slip s_{EM} between the stator field and the rotor. Hence, the synchronous motor runs at a speed that is synchronous to the stator's magnetic field. Since the rotor in most synchronous machines for automotive propulsion includes permanent magnets in the rotor, there is a permanent magnetic rotor field. Therefore, there is no need for an electric connection to the rotor. As a result, the permanent-magnetic synchronous machine offers the same advantage as the asynchronous machine regarding minimized wearing. Additionally, the synchronous speed property and the speed independence of the load over the whole operating range are advantageous features [KWOH17]. The synchronous speed property facilitates simple and precise speed control, which is one key aspect for the application in vehicle propulsion systems [EGLE18, p. 184f]. Furthermore, permanent-magnetic synchronous machines offer high efficiency and a high power density through the use of permanent magnets [GS17].

3.2.2 Storage Systems

The electric energy storage systems in the various voltage levels of the vehicle are the backbone of the supplying infrastructure. They store energy when the power supply is high, for example, caused by the EM in generator mode. If the power demand surpasses the supply, the storage systems release stored energy and balance the power in the APN, ensuring a stable operation. Batteries usually work in a passive manner since they are directly connected to the power network. In contrast, the storage potential in ultracapacitors is actively controlled by a DC/DC-converter, which connects the ultracapacitor with the respective voltage level [Ruf15]. The following sections present the specific characteristics of the primary storage technologies, batteries and ultracapacitors. The characteristics determining the electric behavior and durability have to be considered within the APN management.

Batteries

Batteries are electrochemical storages that convert high-energy reactants into products with lower energy and electricity. In so-called secondary battery cells, this conversion is reversible. Thus, a secondary battery cell can be charged and discharged multiple times. Since the conversion process is not fully reversible, the capacity and other electrical characteristics of the cell diminish, leading to aging during operation and over the lifetime of the battery cell [JBD16, Nea20].

The chemical compounds, as well as the construction technology, determine the battery characteristics, such as specific power, specific energy, or the open circuit voltage (OCV) [GLD02, RLCL14, JBD16]. Other categories for battery characterization are durability, efficiency during charging and discharging processes, or production costs. Another aspect is the dependence in terms of cell temperature, which plays an important role in the development and operation of vehicles in different geographical regions [JBD16, Nea20].

To meet the requirements of the APN, engineers have to make reasonable choices. These choices, such as battery technology, are necessary for every voltage level in a certain power network architecture and have a direct impact on the EMS, the PMS, and the PES. The commonly deployed battery technologies are lead-acid cells for the traditional 12V level and lithium-ion cells for the 42V or 48V level [TT12]. Because of their high energy density, lithium-ion cells are the common choice to supply the electric propulsion system in high voltage levels from 42V up to 800V [BFB+15].

Due to the dependence of the battery aging process on certain factors, such as cell temperature or charging and discharging currents, a battery management system (BMS) constantly measures the battery current and voltage. With these measurements, the BMS determines and monitors the state of the battery storage in terms of SOC and SOH by applying sophisticated algorithms [MOG+09, RLCL14]. Consequently, the BMS protects the battery and ensures durable and safe operation. These measures include limiting charging or discharging currents, compliance with safe voltage levels, and thermal regulation with a thermal management system [RLCL14]. As a result, the battery operation is mostly a compromise between the total operational flexibility and the cells' longevity [MOG+09]. Additionally, the BMS controls the battery balancing procedures. Since individual battery cells vary because of production deviations or aging effects, active or passive cell balancing[4] is deployed [CEH18, CSAH23]. Thus, the full potential of the battery pack is not limited by single cells. Otherwise, the battery pack capacity in a series connection would be limited by the weakest cell with the smallest cell capacity.

In Figure 3.5, the electrical battery behavior is depicted by a combination of a voltage source, resistors, and capacitances [GLD02, GCV19, RSSH21]. The experimental and simulative results in [GCV19] illustrate the dynamic cell voltage V_{cell}, resulting from the electrical battery behavior. The voltage source V_{OC} represents the OCV, which depends on the cell technology and, in particular, on the SOC. In [GLD02], the authors

4 In passive battery balancing, the cell SOCs are balanced by dissipating energy with parallel resistors. In contrast, active balancing approaches transfer electric charge between the cells to achieve balanced SOCs [CEH18].

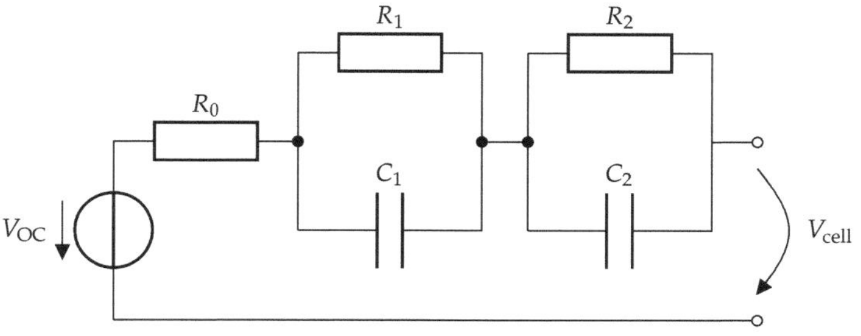

Figure 3.5: Electric circuit diagram of a battery and the resulting cell voltage V_{cell} of the battery cell. The electric elements describe the dynamic behavior of the battery in terms of current and voltage.

show an exemplary OCV graph of a lithium-ion battery and temperature-dependent factors regarding dynamic battery behavior. The resistor R_0 describes the internal resistance, which mainly causes the losses during charging and discharging. These losses in the electric circuit translate into a heat input, which leads to an increasing cell temperature. The following RC-elements, with the parameters R_1, C_1, R_2, and C_2, represent chemical reaction processes inside the battery cell that impact the electrical cell behavior. Since these chemical reaction processes have different transient times, they are described by a respective number of RC-elements, which is related to the model accuracy needed. All parameters and reactions depend on the cell temperature as well as on the SOC, resulting in a highly nonlinear and complex behavior. In summary, all mentioned parameters, and especially the two RC-elements, influence the dynamic battery cell voltage V_{cell} with the time constants

$$\tau_{Batt,1} = R_1 C_1 \tag{3.2}$$

$$\tau_{Batt,2} = R_2 C_2. \tag{3.3}$$

Thus, the battery reacts to arising power demands with an unavoidable delay that depends on the chemical cell reaction dynamics [GCV19]. The time constants $\tau_{Batt,1}$ and $\tau_{Batt,2}$ display the time delay of the battery cell voltage V_{cell} following the electric current.

Batteries are a widely applied electric storage system in the automotive domain. Due to the direct connection to the respective power level, batteries work passively. Thus, the battery pack is charged if the power level voltage is higher than the cell voltage V_{cell} and discharged if the voltage is lower. Therefore, the battery power is a function of the voltage V and the battery states as formulated in (2.2). The individual dynamic response is determined by the cell technology, which is mainly represented by the RC-elements and the respective time constants in (3.2) and (3.3). The static voltage level follows the OCV V_{OC} of the cell, which, in turn, depends on the SOC [GLD02, GCV19].

As a consequence, the proper operation of the battery does not rely on direct control but on reasonable management by the BMS and an adequate voltage level V. The BMS comprises the operational parameters, such as current limitations and thermal control, as well as electric relays to disconnect the battery pack, protecting the battery pack and the APN in terms of serious incidents [RLCL14, CSAH23]. Further, the BMS mitigates long-term wearing and aging effects as it sets adequate current limitations and operation temperatures [MOG+09, RLCL14, JBD16].

Ultracapacitors

The ultracapacitor or supercapacitor offers a higher power density compared to batteries [PPB+07]. Therefore, the use of ultracapacitors adds further flexibility in an HEV or other electric propulsion systems where high electric power is necessary for an effective operation [PPB+07, Ruf15, EGLE18]. Furthermore, the ultracapacitor is a durable storage system that is robust against temperature variations. However, the drawback of the ultracapacitor working principle is the low energy density, which is the object of recent research investigations.

Figure 3.6 describes the physical working principle of an ultracapacitor with an equivalent circuit diagram model. The capacity C_{uc} is charged or discharged during operation and represents the electric storage potential. The ultracapacitor output voltage V_{uc} directly depends on the SOC. The parallel resistance $R_{uc,p}$ for the self-discharging losses and the series resistance $R_{uc,s}$ for the losses during charging and discharging represent the main characteristics. Contrarily to the battery, the ultracapacitor has a single electric time-constant

$$\tau_{uc} = R_{uc,s} C_{uc}, \tag{3.4}$$

which describes the dynamic behavior and is determined by the electric working principle. The time constant τ_{uc} represents the time delay of the voltage following the electric current.

Since the ultracapacitor output voltage V_{uc} directly follows the SOC, the ultracapacitor is mostly connected to the respective power level with a DC/DC-converter making

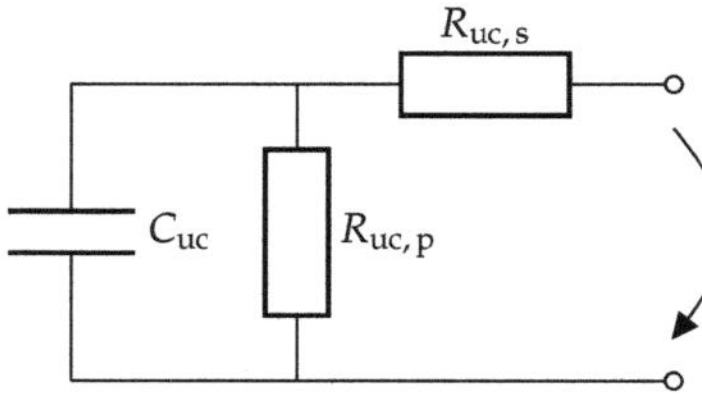

Figure 3.6: Electric circuit diagram of an ultracapacitor with the series and parallel resistance $R_{uc,s}$ and $R_{uc,p}$ and the resulting output voltage V_{uc}. The capacity C_{uc} represents the storage potential of the ultracapacitor.

the whole storage potential accessible [Ruf15]. As a consequence, the ultracapacitor is actively controlled, providing another degree of freedom for energy management [KBH09, KG11a, KG11b, Ruf15]. Hence, in the case of several electric energy storages, the energy management decides how the power demand is distributed and thereby exploits the advantages of the individual storage types [GS17, ORBE20, RSSH21]. In return, the transient response of the ultracapacitor is limited by the DC/DC-converter dynamics [PPB+07, THVY16]. Additionally, the efficiency in operation includes the ultracapacitor losses as well as the DC/DC-converter losses.

3.2.3 DC/DC-Converter

The DC/DC-converter in Figure 3.7, mostly connects different voltage levels in a particular APN architecture (see Figure 3.2 and Figure 3.3). Furthermore, a growing number of components and systems in the APN comprise DC/DC-converters [Ema05, EWK06, Nea20]. For the electrical coupling of two power levels, the converters commonly enable bidirectional power transfer, leading to flexibility in terms of power network operation. To supply individual loads in the APN, unidirectional converters are sufficient since there is only one reasonable power flow direction [KWOH17, p. 103ff].

In order to convert electric power from one voltage level into another voltage level, the DC/DC-converter design comprises energy storage and active switching components. While the active switching components, such as metal-oxide-semiconductor field-effect transistors (MOSFETs) or other transistors, provide the possibility to control the conversion process, the energy storages inside the converter store the electric energy between the switching actions. During switching, the stored electric energy within the DC/DC-converter is released at the desired output voltage level. Since the direct current (DC) at the output results from the switching actions, the DC voltage is only approximated by a high switching frequency. A filter comprising, for example, a capacitance or an inductance is used to get a smoother output voltage.

An exemplary bidirectional DC/DC-converter with capacitors C_1 and C_2 and an inductance L_1 is depicted in Figure 3.7. The time constants between current and voltage caused by these electric elements describe the converter's dynamic behavior. While the inductance L_1 acts as an energy storage and accounts for a delayed current answer, the capacitances C_1 and C_2 act as voltage buffers that smooth short-term power fluctuations [PPB+07, THVY16, KWOH17].

Due to the improvements in recent years and decades, DC/DC-converters with highly efficient MOSFETs and an optimized circuit design offer great flexibility for the automotive power network design [Nea20]. As coupling elements between different power levels, they facilitate the previously mentioned power network architectures with different voltage levels. In turn, these voltage levels provide advantages for the respective loads in the power levels and efficiency. For example, the EM requires a higher input voltage, resulting in a higher efficiency and power density [KWOH17]. Contrarily, most low-power comfort loads were designed for the 12V level and, thus, are optimized for this voltage level [Rei11, Rei14].

Operating as the connection for individual loads, the DC/DC-converter ensures a controllable input voltage level and power flow. Hence, sensitive loads may be controlled individually for a reliable and efficient operation. In [Ema05], the authors even propose APN configurations with converters for every load or small groups of loads. Consequently, these converters contribute to the overall power network controllability by direct load control. Alternative power network designs comparable to the redundant power levels for autonomous driving functions may even comprise several power levels, which are coupled by DC/DC-converters and include their individual energy storages. As a consequence, the respective power levels are independent and more resilient against single points of failure [Kir19].

The aging and wearing effects through exaggerated power transfer and heat accumulation are essential to consider within the DC/DC-converter operation [AK14]. Thus, adequate control and system monitoring are necessary to prevent accelerated wearing of the DC/DC-converter. Consequently, this may result in power transfer limitations in order to mitigate heat accumulation in acute situations [AK14].

3.2.4 Loads – Taxonomy and Description

While storage systems, such as the batteries or ultracapacitors, act as consumers and producers of electric power depending on the APN state and the individual SOC, most loads are solely electric consumers. The respective loads differentiate through the absolute power consumption, the dynamic behavior, or the provided functions. The proposed load taxonomy in Figure 3.8 categorizes load types based on the provided vehicle functions and other properties.

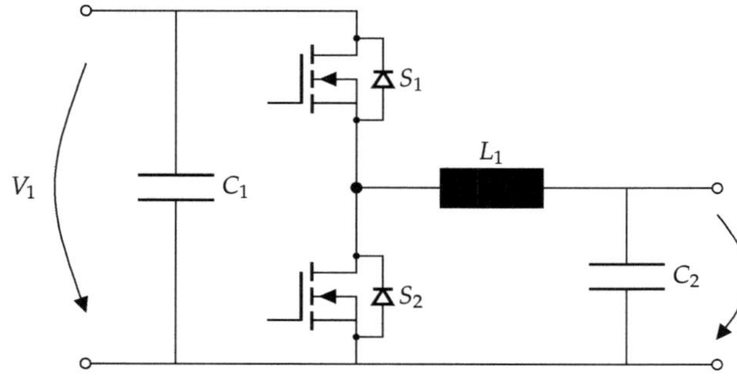

Figure 3.7: Electric circuit diagram of a bidirectional DC/DC-converter with the capacitances C_1 and C_2 and the inductance L_1 as short-term energy buffers. The active switches S_1 and S_2 determine the relation between the voltages V_1 and V_2.

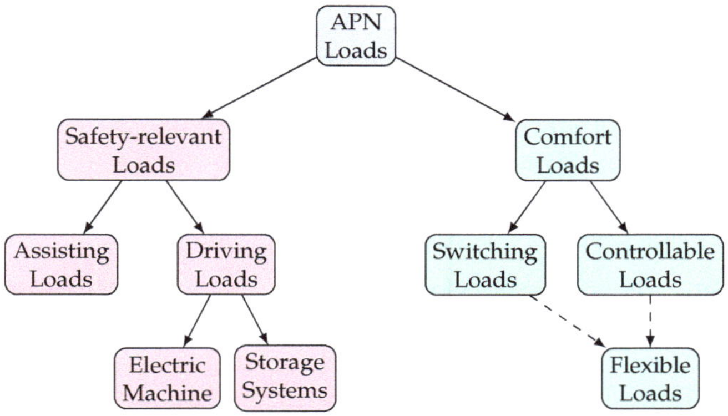

Figure 3.8: Load taxonomy for all loads in the APN. The main categorization concerns the safety or comfort relevance of the loads. Flexible loads are comfort loads with operational flexibility enabling load shaping or shifting.

Power Consumption and Dynamic Behavior

Critical aspects of load differentiation are the average power consumption, the peak power consumption, and the dynamic behavior. The average power demand is a crucial factor for the APN design, the architecture, and the power infrastructure dimensioning. As a result, the HVAC system with a high power demand is located in the 48V level to provide the power with a higher efficiency and to facilitate the application of electric compressors with a 48V input voltage. Further, the storage systems, such as the batteries and the energy sources, need to be in an adequate dimension. The peak power demands, on the other hand, require measures for fast supply adaptation and maybe the supply of higher electric power for a short period. Thus, the EM and the storage systems have to provide certain reserves for short-term adaptations within milliseconds. Finally, the dynamic load behavior poses challenging requirements to the power infrastructure. In this regard, volatility and individual transient times are important factors that have to be considered during the design phase and in the APN management. An example is the electronic suspension system, which causes significant power peaks with high volatility due to the integrated electric pumps and valves. These highly volatile load changes are commonly answered by the DC link capacitor or other capacitances in the APN [PPB+07, Ruf15].

Taxonomy of Load Types

In Figure 3.8, the APN components are distinguished between **safety-relevant loads** in red, on the one hand, and hotel or **comfort loads** in green, on the other hand [REL05]. As a subgroup of the safety-relevant loads, the driving loads comprise the electric powertrain, such as the EM, and the storage systems. Both the electric machine and the storage systems act as consumers and suppliers mainly depending on the APN state, the individual SOC, or the driving mode, for example, in HEVs. These components have a direct influence on the driving operation. Furthermore, the safety-relevant loads include **assisting loads** for the driving operation, such as the electric suspension, the electric steering motor, or the lights. The assisting loads impact vehicle safety and are essential for adequate vehicle operation.

On the other hand, hotel or comfort loads do not contribute to a safe driving operation [REL05]. Therefore, these loads' power reduction or limitation does not compromise vehicle safety in acute situations. However, these loads often play an important role in customer comfort. The comfort loads are further divided into **switching loads** comprising different resistor stages and **controllable loads** operated with PE. The **flexible loads** are another category within the group of comfort loads. Due to their flexibility according to smooth load shifting, flexible loads are notably interesting for intelligent power balancing.

In the following, exemplary electric loads in the APN are described with an emphasis on the general functionality, the power demand, and the dynamic behavior. The loads are presented based on the categories in Figure 3.8 in terms of safety-relevant systems, comfort loads, and flexible loads. For further information and data, it is referred to [Ema05, Rei11, Hes12, Rei14, EGLE18, Nea20] where a comprehensive description of the APN and components and systems is provided. Lists and tables of several electric loads with average and peak power demands are given in [ESG04], in [Büc08, p. 8], in [Rei11, p. 441 and p. 517], in [LR05], and in [Hes12, p. 19ff].

Safety-relevant Loads

In this thesis, the safety-relevant systems include all electric components and systems essential for vehicle safety, ensuring a safe transition to a standstill in an acute and critical situation. This concerns all necessary systems and functions within a time window of about 20 seconds until the vehicle is in a safe state. Hence, in this work, the front or rear windshield heating systems do not belong to this category since their function is safety-relevant for a clear view but not within the respective time frame.

The driving loads, namely EM, and storage systems have been elaborated in previous sections. As the central systems in the electric powertrain responsible for vehicle propulsion, the driving loads have the highest power demands in terms of both average and peak power. For HEVs and EVs, these loads act as power consumers and as power suppliers and build the power infrastructure of the APN.

In contrast, the assisting loads are solely power consumers. The assisting loads provide essential functionalities for a safe driving operation. This load category comprises sensors, computational units, and actuators and covers the whole chain for control decisions regarding autonomous driving functions. The group of sensors includes different systems measuring and observing the internal states of the vehicle and the surrounding environment. Thus, the sensors facilitate adequate control and management of the internal systems on the one side and set the foundation for autonomous driving decisions on the other side. Exemplary sensors are current and voltage sensors in the APN, temperature sensors for components and the interior, cameras, ultrasonic sensors, radar sensors, and lidar sensors. Since the sensors continuously measure and provide information, the power demand is constant during driving. Hence, the peak and the average power demand are comparable.

For the group of computational units, the instantaneous power demand varies in relation to the computational load. In particular, this is true for the units with higher computational power, which host software functions that may vary with the objects around the vehicle, the driving situation, or the computational demand from telecommunication functions. On the contrary, most ECUs provide highly integrated and adapted functions that cause a constant power demand. However, with the emergence of zonal architectures and the centralization of software functionalities, the number of centralized computational units in the vehicle increases, resulting in a volatile power demand.

Actuators for autonomous driving functions pose the highest average and peak power demands. These often electromechanical components and systems affect physical actions, such as the manipulation of the steering system, and have to overcome a certain friction moment. Therefore, most actuators are electric motors comparable to the EM for the vehicle propulsion and comprise a fast dynamic behavior and volatile power demand. In contrast to the EM with its often bidirectional inverter, they only consume power. One exceptional actuator with an enormous peak power demand is the electric suspension system [Nea20, p. 107ff]. The electric suspension affects the vehicle height as well as the suspension characteristics and thereby directly enhances the vehicle's controllability and driving stability. Since these measures need precise and discontinuous actions, the power consumption differs over time. Consequently, the average power demand for the various electric suspension types is low compared to the peak powers. This refers, in particular, to critical driving situations, such as steep turns or evasive maneuvers. Other examples are the anti-lock braking system (ABS) or the electronic stability program (ESP), which cause precise and intermittent interferences, leading to a volatile and unpredictable power demand.

Comfort Loads

In contrast to the safety-relevant systems, the comfort systems comprise all loads, which are not essential in an acute and critical driving situation. Thus, they may restrict functionality without compromising the vehicle safety and stability. However, some systems in this group have important auxiliary functions, such as the thermal management of the engine or the front windshield heating for a clear sight in winter.

In addition, the hotel or comfort loads usually directly determine the passenger comfort in the car. Since customer comfort should be met at any time, a noticeable load reduction is the last measure in order to stabilize the APN voltage levels. For further differentiation, the comfort loads are categorized according to their controllability. Therefore, the comfort loads are distinguished into switching loads and continuously controllable loads.

The on-off switching loads comprise mostly traditional components with a resistor or other electric components that are not continuously controllable. This subgroup usually includes interior lights, the radio, and telecommunication and entertainment systems, such as speakers or displays. Furthermore, the load group comprises electric seat adjustment motors, electric window regulators, or the electric sunroof. As a result, the passengers directly notice the manipulation if a load is automatically switched on or off. One example is the heating resistor within the thermal management that facilitates additional heating at the vehicle start or in cold conditions and is not controllable by the passengers. In general, the loads in this subgroup have a constant power demand while activated. The activation or deactivation affects an inevitable rise or fall concerning the electric power. The respective steepness depends on the load's electric design regarding internal capacitances or inductances.

Compared to the on-off switching loads, the stepwise switching loads offer more control actions than activation or deactivation. This is due to the application of several resistors, which switch with regard to the intended function or discrete power levels within the system. A prominent load in this subgroup is seat heating, which mostly includes resistors for several power stages. With every connected resistor, the heating power increases, and the power demand according to the APN changes. Other components may be the blowers, the windshield heating, the wipers, or the driving wheel heating system. The dynamic behavior is comparable to the previously described on-off switching loads. However, because of the stepwise switching, the respective power rises or falls may be higher or lower, resulting in a particular impact on the power supply and voltage stability.

The last subgroup includes all comfort loads and systems that comprise the possibility for continuous control, for example, by integrated power electronics (PE). A system representing a significant power demand in most APNs is the HVAC system, which controls the vehicle's heating, cooling, and air conditioning. The HVAC system conditions the vehicle interior according to passenger comfort. As a central component of the HVAC system, the electric compressor includes PE to control its functionality smoothly. The battery thermal control may also be coupled with the HVAC cooling circuits. Other components and systems in this subcategory are some blowers, overall thermal management, motor fans, or pumps. Most components constitute connections between the electrical system and mechanical parts in the vehicle and include motors or other electromechanical actuators. Due to the progressive electrification of former mechanical-propelled actuators, this subgroup has a growing number of components.

With regard to the dynamic electric behavior, these loads offer the most flexible controllability resulting from the integrated PE. As a consequence, the power demands are smoothly adaptable following the available power and the intended function.

Flexible Loads

A highly relevant load property is operational flexibility, which, in this thesis, is defined as the load feature to shift or shape the instantaneous power demand without compromising the intended function. The operational flexibility is either given if the function is not time-critical or if there are energy storages within the respective system that cause some inertia.

The most prominent system in this category is the HVAC system, which includes multiple inherent energy storages. Some examples are the coolant liquid inside the cooling circuit or the air and the surfaces in the vehicle interior. Both are inherent energy storages and represent a respective capacity for heat, causing a specific time constant for heating or cooling the system. On the one hand, this affects the slow down of the control actions since, for example, the introduced heat will be integrated to subsequently change the temperature. On the other hand, in case of shifted or reduced control actions, the temperature storages offset the missing control actions to some extent. Therefore, the electric compressor operation may be shifted without the passengers noticing.

However, the acceptable temperature range for the specific heat storage is important in determining the heat storage potential. According to the vehicle interior, the temperature should remain within a range of $\pm 1\,^{\circ}\mathrm{C}$ lower or higher than the desired temperature. This is because the passengers do not recognize this slight deviation [LLDL11]. For other storages, the range is determined by the technical design and requirements for proper function.

In summary, the internal storage capabilities allow storing thermal energy and, by exploiting these storages, the HVAC system may shift or shape power demands without compromising its functionality in the short term. Regarding the power balancing in the APN, operational flexibility is a valuable property that adds beneficial opportunities. Other systems in the APN, which provide this feature, are the seat heating or the thermal management of the powertrain and other components. Some research works propose the usage of these flexible loads to shift demands and thereby mitigate power peaks in the APN [LLRB+14, VA15] or optimize vehicle efficiency [ÅEB+04, KKJ+07].

3.3 Communication Technologies

One factor concerning the APN management is the applied communication technology in both dimensions, hardware, concerning wiring, and software in terms of communication protocols. This section describes the various automotive communication technologies regarding their features and properties. The description mostly follows [Rei11, TGH+15, RGKS20, ORBE20, ZZL+21]. A concise comparison between the classic signal-based communication and the communication based on the SOA is provided in Table 3.1.

Established Automotive Communication Technologies

In the beginning, the communication in the vehicle was signal-oriented through the deployed hardware since point-to-point wires facilitated communication. In 1991, the controller area network (CAN) bus was the first communication bus system introduced to a series vehicle [Rei11]. As a bus, it aggregates the individual connections between the respective components and systems in the vehicle, resulting in a simplified communication architecture with equal communication nodes but reduced physical cables in the car. Thus, the CAN bus introduction provided a lot of advantages compared to the direct connection between all components [Rei11]. The number of nodes within a CAN network is theoretically unlimited. However, the constrained data volume and the physical characteristics in terms of signal transmission restrict the number of nodes [Rei11, p. 105]. The CAN technology differs in classes with different speeds and data transmission, mainly the high speed with up to 1 MBit/s and the low-speed implementation with up to 125 kBit/s, which are adapted for the communication needs in specific applications and vehicle domains [Rei11, p. 105] [ORBE20]. While the high speed CAN connects systems of the powertrain, which depend on responsive communication for capable control, the low speed CAN bus builds the communication backbone for the less dynamic comfort loads, such as the HVAC system, the electric sunroof, or the lights [Rei11]. The technical definition of the CAN standard allows prioritization of the several nodes that are interconnected through the bus system. With the increasing number of electric components and computational units (ECUs) in the vehicle, the number of CAN buses and the amount of data steadily grows [ORBE20].

Since the increasing number of CAN buses in the vehicle implies additional cost and weight, the automotive industry introduced the local interconnect network (LIN) standard to relieve the CAN communication and thereby reduce production costs [Rei11]. With the additional LIN bus in the overall communication architecture, there is a cheap possibility to connect up to 16 components in a local network with a data transmission of up to 20 kBit/s. The communication is driven by a master node which connects the network to a superordinate ECU or a CAN bus and triggers the slave nodes in the network [Rei11, ORBE20].

Today's vehicle communication architecture is highly adjusted and capable. It comprises several CAN and connected LIN buses to the numerous components in the APN [TGH+15]. However, upcoming requirements for growing data transfer have led to additional technologies in recent years that augment the communication structure. One example is the implementation of the media oriented systems transport (MOST) bus, which is highly adapted to connect infotainment and telecommunication systems. In the MOST standard, the communication network supports up to 64 systems in the bus and a data rate of 25 MBit/s (MOST25) or even more in recent updates (MOST50 and MOST150) [TGH+15, ORBE20]. Two further technologies, complementing the traditional automotive CAN and LIN architectures, are FlexRay and controller area network flexible data-rate (CAN FD) [Rei11, ORBE20]. Both communication standards improve the data rate and provide a more reliable and deterministic communication than the MOST bus.

Automotive Ethernet and Service-Oriented Architecture

The trend for more infotainment systems and autonomous driving functions demands higher and more flexible data rates. Therefore, the in-vehicle communication architectures are enhanced by the introduction of Automotive Ethernet [HMVv13, TGH+15, ORBE20]. Contrarily to LIN or CAN, the Ethernet technology was not developed for the specific application in vehicles or neighboring domains but for the general use in broad networks within buildings and around the world [LVH11]. Hence, Ethernet brings scalability and flexibility for future in-vehicle networks [HMVv13, HSS+22].

Table 3.1: Comparison of the different properties and features of recent automotive communication technologies (Classic) and the SOA with reference to [Rei11, TGH+15, RGKS20, ORBE20, SSG+22].

Category	Classic	SOA
Communication Patterns	Signal-oriented	Publish-subscribe, request-response, fire-forget
Path Configuration	Predefined at vehicle production by C-Matrix	At vehicle start or at runtime with SOME/IP-SD
Hardware and Bandwidth	LIN (Single wire, 20kb/s), CAN (Twisted pair, 1 Mb/s), MOST (Optical fiber, 150 Mb/s), CAN-FD (Twisted pair, 10 Mb/s), FlexRay (Twisted pair or optical fiber, 20 Mb/s)	Automotive Ethernet (Twisted pair, 1 Gb/s)
Strengths	Well-established, tailored hardware, secure	Flexible design, flexible operation, high bandwidth
Weaknesses	High efforts in development, inflexible after start of production	High cost, complex regarding communication patterns

Due to the implementation of different hardware, the bandwidth ranges from 10 MBit/s up to 1 GBit/s, which qualifies this communication technology for the application within vehicle domains or as the communication backbone between all domains [LVH11, ORBE20]. Since Automotive Ethernet is able to provide a high data rate and flexible communication paradigms, it is a possible technology to enable the broad application of autonomous driving functions. However, the required robustness and fail-safety in this domain calls for further developments and standardization [HMVv13, HSS+22, SGLS22]. Recent standards referring to Automotive Ethernet comprise definitions on the hardware as well as on the software level to improve robustness.

In comparison to the previous technologies, the combination of Automotive Ethernet, on the hardware side, and the service-oriented architecture (SOA), on the software side, facilitate a flexible in-vehicle communication [RGKS20, HSS+22, SSG+22]. The traditional automotive communication design predefines the transmission paths by a signal-based structure described in a communication matrix (C-Matrix) [HSS+22]. With the introduction of Ethernet and the SOA, there is no need to define the interconnection between components in advance since Ethernet protocols provide a flexible interaction buildup during runtime [SSG+22].

Whereas the recent software design in cars is a tight coupling between the ECUs and the specific software functions, future designs based on the SOA enable the division of hardware and software, leading to enormous flexibility [SGLS22]. Respective software applications may be instantiated on multiple ECUs to increase safety in case of single ECU failure or to distribute the computational effort [SGLS22]. The arising flexibility with Automotive Ethernet and the SOA is a crucial factor for future developments in the automotive domain and, particularly the flexibility required for APN management.

As elaborated in [RGKS20], the automotive SOA differentiates from traditional communication schemes in terms of new patterns and principles. In contrast to the signal-based bus communication, which is predefined by the communication matrix, the SOA comprises the three flexible patterns, publish-subscribe, request-response, and fire-forget [HSS+22]. In the SOA context, communication always occurs between the client asking for information or a service and the server providing the information or the respective service [RGKS20].

Regarding communication patterns, publish-subscribe replaces the traditional signal-oriented communication. However, the publish-subscribe pattern allows for the subscription at vehicle start or even during operation while the signal-oriented communication is defined before vehicle production in the communication matrix [HSS+22]. With the subscription, the client receives regular information from the server, triggered by events or on a periodical basis. On the other hand, request-response enables a one-time information query or the service invoking on the server. Moreover, with the fire-forget pattern, the client sends information or commands to the server without requesting an explicit answer, for example, in terms of control commands.

The described communication patterns are defined in different specifications and protocols, such as scalable service-oriented middleware over IP (SOME/IP) and data distribution service (DDS), which are standardized for the automotive industry in AUTOSAR [RGKS20, HSS+22]. The fundamental communication principles emphasize a data-centric design and the communication path configuration at vehicle startup or runtime. To facilitate the configuration of new communication paths, the SOME/IP Service Discovery (SOME/IP-SD) provides respective protocols for clients to find and subscribe to information or services in the network. On the other hand, there is the possibility for the servers to publish their information and services. Hence, through the communication patterns and service discovery mechanisms, the automotive SOA allows for great flexibility of modern in-vehicle communication [SSG+22].

3.4 Goals Toward Modern Automotive Power Networks

This section elaborates on the main goals concerning the APN design and operation of modern and future vehicles. Most goals are derived from the previous descriptions and the three major trends in Chapter 1, namely electrification, dissemination of telecommunication and entertainment systems, and autonomous driving functions. The described goals display different perspectives on the APN and concern the automotive power management to varying extents.

Consequently, the following goals for the APN are discussed from the two perspectives of hardware and software:

- ◇ Quality of power supply and voltage stabilization
- ◇ Flexibility for customization and personalization
- ◇ Fail-safety of APN operation (for autonomous driving functions)
- ◇ Electric efficiency of APN operation
- ◇ Reduction of costs, weight, and resources
- ◇ Increased durability and longevity for batteries and other components

Quality of Power Supply

The first and foremost goal of the APN is a reliable and stable power supply. According to this goal, the tasks are

1) to ensure a stable voltage level in all APN voltage levels,

2) to guarantee the availability of electric power during power peaks,
3) and to provide enough energy for all vehicle functions.

Consequently, there are definitions for voltage ranges and transients that have to be guaranteed in the different voltage levels [ERWL05, Büc08]. Nevertheless, to ensure the required voltage stability, the power infrastructure, comprising the EM in generator mode or the electric storage systems, has to provide the necessary amount of electric power, which becomes difficult through the current developments and trends [KWT+10, Hoh10, TVE+22]. Additionally, in the long term, these supplying systems have to deliver the necessary energy to deploy all demanded vehicle functions.

Hence, the APN infrastructure has to meet the short-term requirements in terms of load transients and peak power demands as well as the long-term requirements for the average power demand. An important hardware element concerning the transient behavior of the APN are capacitances within the power infrastructure [Ruf15]. As an example, the DC link capacitor of the DC/DC-converter filters the output voltage and provides highly responsive energy storage to answer sudden load changes [PPB+07, THVY16]. In case of a short circuit, the APN harness comprises different fuses to protect the components and the power supply for the remaining areas. With the ongoing development in semiconductor technology, these fuses and relays become increasingly intelligent and facilitate additional opportunities, such as switch-off with the possibility for reconnection [GFKB22].

The reliability of the power infrastructure mostly builds on the three pillars: energy availability, adaptable power supply, and voltage stability through the underlying control loops. These pillars correspond to the three layers, energy management shell (EMS), power management shell (PMS), and power electronics shell (PES) in Figure 2.2. The most researched software parts and the foundation for a stable voltage in the different power levels are the underlying control loops concerning the EM or the DC/DC-converter [THVY16, KWOH17]. These are highly individual and tailor-made since they control various components and systems in the APN power infrastructure in distributed manner [Gem15]. According to the reliable power supply and a stable voltage level, the key control algorithms are the EM operation, including the inverter, as well as the DC/DC-converter control. Both components are active elements in the power infrastructure and represent the main power sources strongly affecting the supplied voltage level. The dynamic behavior of the DC/DC-converter affects two voltage levels since it is a power sink on one level and a power source on the other. Due to the respective time constants within some milliseconds, the control loops for the EM inverter and the DC/DC-converter need to be responsive, facilitating an adequate system performance. Thus, some parts of the control algorithms, in particular, for the DC/DC-converter, are directly implemented in hardware elements to increase the responsiveness [THVY16]. In contrast, the complex dynamic behavior of the EM requires a cascaded control design or the application of advanced control algorithms, for instance, an MPC approach [Gem15, EGLE18].

In contrast to the underlying control loops, the coordination and management levels operate with a longer cycle time. By coordinating and distributing the energy and power supply, the algorithms at this level ensure the suitable deployment of the different storage systems [ABR+10]. In this regard, the general task is to achieve a reasonable power or energy balance. Besides the underlying control loops, which have to be more capable in future vehicles, the coordination and management levels are confronted with significant challenges. This is mainly caused by the growing number of components in the APN resulting in more power levels, an increased power demand, and coordination effort. Especially, the exaggerating power demand in the APN, concerning both the average and the peak power, poses noteworthy challenges to automotive power management [Hes12, p. 21], [KWT+10, RWM+13].

Therefore, intelligent power management approaches need to utilize the different power sources and reasonable load management to suitably coordinate the power distribution within the APN. One way to achieve better power management in this context is to predict future loads regarding the driving situation or the passengers' behavior [Kir19, GLS+20].

Flexibility

Another aspect caused by the possibilities of customization is the enormous number of different variants and models, which have to be achieved by the hardware and software design. In order to manage the reasonable development and production of new vehicles, flexibility and modularity are greatly needed. According to the hardware side, this challenge drives the trend toward, for example, a zonal architecture that allows for a seamless introduction of new components and systems in the respective zones during the development process, during production, and even after the production [RWKT13, Jia19, ORBE20, ZZL+21].

Furthermore, a zonal architecture aggregates the computational power in a limited number of decentralized computation units where most of the data will be processed [WG20]. The power distribution is another aspect that may change from a star to a tree design with the emergence of a zonal architecture. In [TTFH18], the authors even propose a ring structure for the core power infrastructure, resulting in a reliable power supply. However, the individual zonal controller has to provide enough plug-in connections, offering the desired flexibility for vehicles. As a result, this development affects the vehicle's power supply, data transfer, communication, and computational power.

On the software side, customization and personalization lead to more functions in the vehicle, ranging from telecommunication and entertainment to autonomous driving. The individual and distributed software parts, as well as the management levels, have to deal with a sophisticated interconnection of systems [SSG+22]. Only the broad use of a flexible and modular algorithmic design can answer the current trends and facilitate a well-functioning software update over the air regarding all functions in the vehicle [HSS+22]. In this regard, automotive power management is one part of the system that needs to become flexible and modular to support the plug-and-play integration of new hardware components. Additionally, the increasing interplay of systems demands more modularity and flexibility in terms of communication technologies and protocols [RGKS20, SSG+22].

Fail-safety

The need for fail-safe operation grows with the ongoing dissemination of autonomous driving functions and the electrification of former mechanically propelled safety-relevant systems. Since the trend toward autonomous driving functions removes the driver as the safety fallback, the requirements for functional fail-safety increase and affect the power infrastructure as the basis for vehicle operation [WSD+18, Jia19]. All components and systems of the autonomous driving functions, such as sensors, computational units, and actuators, rely on a stable power supply [Kir19]. Furthermore, through the ongoing electrification, there are safety-relevant components, such as the brake servo or the steer-by-wire and brake-by-wire systems, which are supplied by the electric power infrastructure as the primary energy source. Thus, the APN reliability directly relates to vehicle and passenger safety. Therefore, measures, such as redundant fallback power levels, become essential in vehicle concepts [WFF+16, GKO19, TTFH19].

At the same time, fail-safe software is crucial for vehicle safety and introducing autonomous driving functions. Since the electric power infrastructure is the backbone for safety measures and autonomous driving functions, its control and coordination through power management have to fulfill the exact safety requirements as the hardware components [WSD+18, Jia19]. As a result, vehicles comprise redundancy in essential software parts in order to facilitate a suitable fallback solution in case of failures [SGLS22]. Consequently, the goal of fail-safety concerns automotive power management in various aspects.

Electric Efficiency

The electric efficiency of the APN strongly relies on the individual components and systems as well as on the power distribution system. With the ongoing electrification in today's vehicle concepts, such as HEVs and EVs, there is a growing interest in component efficiency. Electric power and energy are no longer a byproduct the alternator provides but gain value while determining the driving range. Thus, the power infrastructure and automotive power management have to become increasingly efficient. This concerns the resistive and switching losses during operation as well as the reduction of power consumption during standstill [Dör16]. In this regard, a contradicting aspect is the required availability of vehicles. Whether it is the possibility for updates over the air or the general availability for the user who wants to examine the vehicle state via a mobile app before the ride, some parts of the vehicle may have to be available even during longer standstills [WSD+18, ZZL+21, HSS+22].

Another aspect of efficiency is the propulsion system and, for example, the possibility for regenerative braking, which, in turn, poses requirements to the APN design [HWK+17, GS17]. The intelligent and extensive use of regenerative braking is a major factor in efficient driving operation and may be enhanced by capable power management during recuperation. In general, efficiency plays a vital role in automotive software concepts. This concerns the efficient operation of the APN by the applied management approaches as well as the software itself in terms of computational efficiency. With reference to the proposed software architecture for automotive management shells in Figure 2.2, energy efficiency is one of the primary energy management tasks. In contrast, the automotive power management focus lies on load balancing.

Costs, Weight, and Resources

Since the mass production of vehicles includes high numbers of the same components and systems, these systems have to be as cheap as possible. This concerns the production costs as well as material costs. Another aspect, in this regard, is the sustainability and resource efficiency during production with an emphasis on the vehicle lifetime. With a long-term and expandable vehicle platform, there is a chance for more sustainability. A vehicle platform that can be complemented with additional or updated hardware, such as telecommunication and entertainment systems, will likely be used longer after production. Accordingly, power management approaches facilitating plug-and-play integration contribute to a sustainable vehicle platform.

Further, weight reduction is an important aspect that has implications for safety and overall driving efficiency. Therefore, hardware dimensioning is always a trade-off between proper operation and weight constraints. Regarding battery dimension and capability, this trade-off requests capable and effective automotive power management in terms of continuous power balancing [KWT+10, RWM+13].

Durability

An important aspect that is connected to cost reduction and sustainability is the longevity for all components and systems in the APN and for the battery, in particular. Since the requirements regarding the battery and other storage systems rise, there is a need for durable battery technologies and proper and reasonable operation [MOG+09]. Due to the required cost and weight reduction, there is always a trade-off between a sufficient battery size leading to a reduction in wearing effects, on the one hand, and the minimal dimensioning for reduced weight, resources, and costs, on the other hand [Nea20]. However, battery storage with sufficient reserves is also beneficial toward a reliable and stable power supply and, thus, has implications for vehicle safety [Ruf15, EGLE18]. Hence, this goal constrains the possibilities of power distribution and adds further relevance to the automotive power management coordination task. In this regard, power management has to compensate for the reduced battery size by a frequent power balancing.

3.5 Conclusion

This chapter elaborates on the fundamentals of automotive power networks and creates the foundation for the proposed predictive auction-based power management in the next chapter.

First, the different voltage levels and their evolution as a major decision for a suitable power network architecture are described in Section 3.1. Secondly, in Section 3.2, the main components in the APN, such as the EM, the DC/DC-converter, and the electric storage systems, are introduced. The focus is on the electric behavior and, in particular, on the time constants that influence the electric dynamics in the different voltage levels. As depicted in Figure 2.2, these APN dynamics and the PES are directly interconnected with the automotive power management and its power balancing task. The automotive power management in the next chapter is developed with respect to the described component characteristics and behavior.

The load descriptions and taxonomy in Section 3.2.4 build the foundation for coordinated load reduction through automotive power management [KFB+10, RBW+12]. Accordingly, the section lists several comfort loads that are appropriate for load reduction in order to achieve a power balance in the APN.

With regard to the requested flexibility and modularity, the presented communication technologies in Section 3.3, with an emphasis on Automotive Ethernet, indicate a promising evolution in the automotive industry [SGLS22]. Finally, Section 3.4 formulates goals toward modern APNs that correspond with the current trends. In accordance with the identified research gap, one necessary step to achieve these goals is a flexible[5] and capable automotive power management which is investigated and presented in the remainder of this work.

[5] The flexibility of the automotive power management mainly concerns the desired plug-and-play integration of new hardware components into the APN during development and even after production as defined in Section 2.2.

4 Auction-based Automotive Power Management with Predictive Extension

This chapter presents the investigation and development of auction-based automotive power management to close the research gap specified in Section 2.6. In Section 4.1, the uniform price auction (UPA) is introduced and mathematically formalized as a power balancing mechanism. Subsequently, the UPA is transferred to the application in automotive power management and extended for power networks with multiple voltage levels. Moreover, a hysteresis mechanism is proposed to integrate switching loads into the UPA.

The predictive extension that increases the power balancing capability by exploiting flexible loads is introduced in Section 4.2. The extension exploits the HVAC system as a flexible load, which provides the possibility to shift future power demands without compromising user comfort.

The UPA framework facilitates the seamless integration of new loads into the power balancing mechanism. The proposed auction-based power management is complemented by the flexible and modular communication design based on the SOA in Section 4.3 to achieve the desired plug-and-play integration property. The conclusion in Section 4.4 summarizes the key accomplishments.

4.1 Modern Auction-based Automotive Power Management

As highlighted in Section 2.1, the first goal of the APN is a reliable and stable power and energy supply. However, this is a challenging goal due to the rising number of electric loads with a growing power consumption, peak power, and highly dynamic power transients. In this regard, the power management task is balancing power supply and demand in the APN during phases of power imbalances.

4.1.1 Formalization of the Auction Mechanism

A promising methodology facilitating the required power balancing and the desired plug-and-play integration is the auction-based approach first presented in [Gra04a] and patented in [Gra04b]. In [Büc08], the approach is adapted and explained in detail regarding the mathematical background of auction theory and the corresponding economic sciences. For the fundamentals, theoretical insights, and practical examples of auction theory, it is referred to the works in [Kle04, LM05, MBC11, NZ20]. The basic approach for auction-based power management in [Gra04b] has been extended for vehicles with multiple voltage levels in [SRSH21, SSH21] and formalized in [SGH24].

Formalization of the Uniform Price Auction in the Energy Market

Auctions in various forms played an important role in the history of humankind [Kle04]. The most known and common auction mechanism is the open auction with a steadily ascending price for a particular item. However, with reference to the traded goods or items, the participants, and the circumstances, there are various auction mechanisms for a broad range of applications. Today's electricity markets in different countries and continents are one successful example with regard to the practical implementation of auctions [MBC11, NZ20].

Besides other auction forms, the uniform price auction (UPA) is a commonly implemented auction type in the energy market [Fri60, NZ20]. The main characteristic of the UPA is the determination of a uniform price p for a continuous good, for example electric power[6], offered by the sellers and demanded from the bidders as visualized in Figure 4.1. In this regard, the UPA represents a market with offer and demand. The uniform price p is called the market clearing price (MCP). The central auctioneer

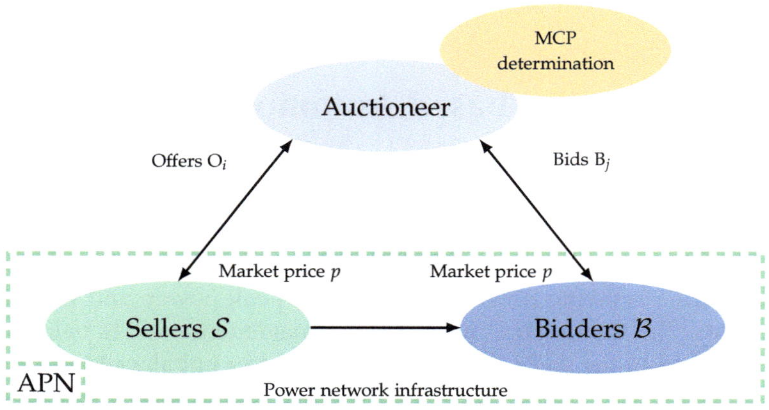

Figure 4.1: Schematic of the interaction within the UPA in the context of power balancing in the APN. The auctioneer compares the offers O_i and bids B_j to determine the market clearing price (MCP), which depicts the market price for the balance between demand and supply [SGH24].

[6] In the following, the UPA is utilized for the balancing of electric power supply and demand. Nevertheless, the auction mechanism is applicable for all continuously dividable goods.

compares the offers O_i and bids B_j to determine the MCP. The MCP is defined as the price p for which the amount of offered and demanded electric power is equal [NZ20]. Hence, with the MCP, the energy market is cleared according to the electric power for the next time interval. Therefore, the existence of a unique MCP, shown in Figure 4.2a, is necessary to achieve the desired power balance. For the energy market, the UPA takes place continuously to coordinate the consumption and supply for the next period in the proceeding time horizon [MBC11, NZ20]. After the auction, the sellers and bidders exchange the electric power according to the determined price p. The actual MCP determination relies on the respective offers and bids. The steps for a single UPA cycle are depicted in Figure 4.3 while the detailed UPA description is given in Definition 4.1. The following definitions and the formalization of the UPA have been presented in [SGH24].

The graphs of offers and bids in Figure 4.2 demonstrate the UPA working principle. The offers and bids in Figure 4.2a are conform to the proposed rules in Definition 4.1. Hence, there is an intersection between the aggregated offers and the aggregated bids, representing the MCP. However, in the other market situations, the UPA does not lead to the market clearing or the exchange of electric power since no MCP exists. In Figure 4.2b, the offers and the bids are not defined for the whole market price range. On the other hand, in Figure 4.2c and Figure 4.2d the third rule in (4.6) with $O_i(p_{\min}) = 0$ and $B_j(p_{\max}) = 0$ is not met.

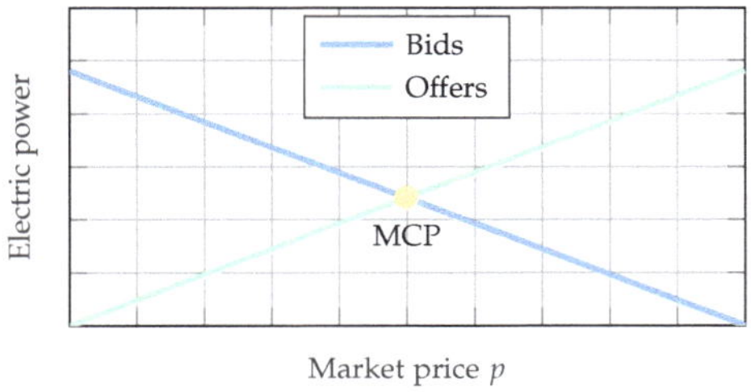

(a) MCP as the intersection between bids and offers.

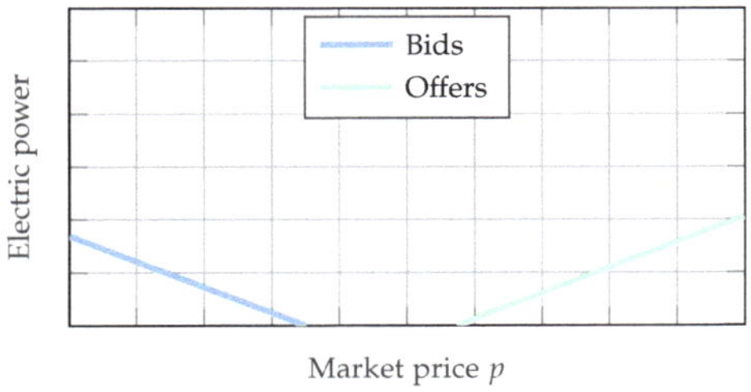

(b) No intersection between bids and offers.

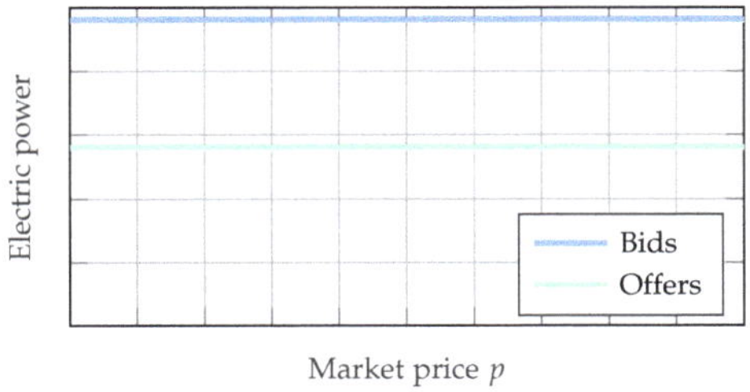

(c) No intersection due to constant offers and bids.

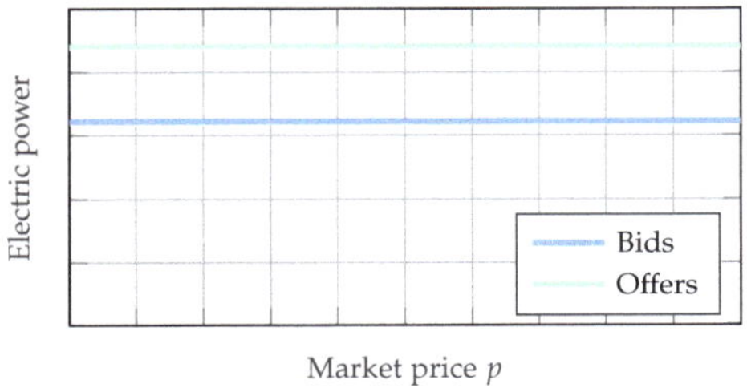

(d) No intersection due to constant offers and bids.

Figure 4.2: Different constellations of bids and offers within a UPA concerning the general working principle and possible edge cases that impair the MCP determination and, thereby, the market clearing.

Definition 4.1: Uniform Price Auction [SGH24]
The uniform price auction comprises a group of sellers

$$\mathcal{S} = \{1, 2, \ldots, s\},\ |\mathcal{S}| = s \geq 1 \tag{4.1}$$

and a group of bidders

$$\mathcal{B} = \{1, 2, \ldots, b\},\ |\mathcal{B}| = b \geq 1, \tag{4.2}$$

which exchange electric power for a certain market price p. To bargain the amount and the price p of electric power, the sellers $\mathcal{S}$ and the bidders $\mathcal{B}$ propose offers O_i with $i \in \mathcal{S}$ and bids B_j with $j \in \mathcal{B}$ depending on the market price p.
The central auctioneer compares the offers O_i and bids B_j for electric power to determine a uniform market price, which is called the market clearing price (MCP):

$$\sum_{i \in \mathcal{S}}^{s} O_i(p) = \sum_{j \in \mathcal{B}}^{b} B_j(p) \text{ with } MCP = p. \tag{4.3}$$

The existence of a unique MCP is necessary to achieve the described power balance in (4.3). To guarantee a unique MCP, there are three rules concerning offers and bids:

R1: Offers O_i are monotonically increasing for an increasing market price p:

$$O_i(p_1) \geq O_i(p_2) \text{ for } p_1 > p_2 \text{ and } i \in \mathcal{S}. \tag{4.4}$$

R2: Bids B_j are monotonically decreasing for an increasing market price p:

$$B_j(p_1) \leq B_j(p_2) \text{ for } p_1 > p_2 \text{ and } j \in \mathcal{B}. \tag{4.5}$$

R3: The offers O_i and bids B_j are specified for the whole market price range $p \in [p_{min}, p_{max}]$. For the boundaries of the price range, the following assumptions apply:

$$O_i(p_{min}) = 0 \text{ and } B_j(p_{max}) = 0 \text{ for } \forall i \in \mathcal{S}, \forall j \in \mathcal{B}. \tag{4.6}$$

After the uniform price auction, the sellers $\mathcal{S}$ and the bidders $\mathcal{B}$ provide or consume electric power according to the determined MCP and the offers O_i and bids B_j.

With the claim in the three rules, an intersection within the valid price range $p \in [p_{\text{min}}, p_{\text{max}}]$ is guaranteed. Since all bids B_j are zero for the maximum market price p_{max}, all offers O_i are zero for the minimum market price p_{min}, and with the rules in (4.4) and (4.5), it exists at least a market clearing for $p = p_{\text{min}}$ or $p = p_{\text{max}}$. In both cases, the traded electric power is zero because of no demand for $p = p_{\text{max}}$ and no supply for $p = p_{\text{min}}$, respectively. Figure 4.3 visualizes the auction procedure formalized in Definition 4.1.

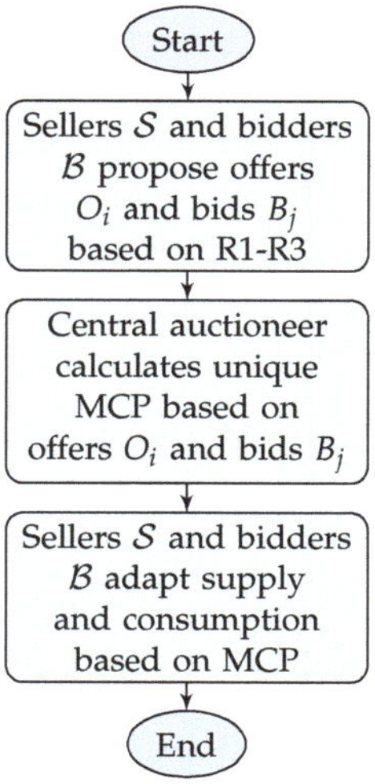

Figure 4.3: Procedure of the basic UPA summarizing the steps in Definition 4.1 for one auction cycle.

Uniform Price Auction in the Automotive Power Network

Since the energy market and the APN have similar architectures and tasks, some mechanisms suit both domains. In [Gra04a], the UPA has been transferred from the energy market to the automotive power management for conventional vehicles with a single 12V power level. Since the power management task in [Gra04a] is applied with a cycle time of 200 ms, the auction algorithm is executed more frequently compared to the energy market, where the UPA is utilized to negotiate the energy supply and consumption for the next minutes, hours, or days [MBC11, NZ20].

The automotive power management coordinates the electric power for the next time interval. Therefore, the central auctioneer in Figure 4.1 compares the offers and bids for power and determines the MCP to clear the market and thereby achieve a power balance in the APN. After every auction, the electric power is transferred from the sellers $\mathcal{S}$ to the bidders $\mathcal{B}$ via the power network infrastructure. The amount of electric power and, thus, the consumption and supply of the components is determined by the current MCP. To ensure the existence of an intersection between the power supply and demand, the UPA mechanism has to meet the proposed rules in Definition 4.1.

In this work, the market price range in the third rule is arbitrarily specified to

$$p \in [0, 10], \tag{4.7}$$

based on the concept in [Gra04a][7]. In the following, PPFs as described in Definition 4.2 are utilized to express the offers and bids.

[7] In the remainder of the thesis, the price interval $p \in [0, 10]$ is mainly used for simplification. However, the price interval is a design variable for the general UPA described in Definition 4.1.

The PPF in Definition 4.2 assigns exactly one electric power value of P to every market price p in the valid price range. Additionally, the PPF_i of a component $i \in \text{APN}$ comprises parameters that correspond to the respective component i. The resulting graphs for an exemplary APN are visualized in Figure 4.4. The PPF_{bids} aggregates all bids B_j of the bidders while $\text{PPF}_{\text{offers}}$ aggregates all offers O_i of the sellers. Due to the compliant design of PPF_{bids} and $\text{PPF}_{\text{offers}}$, there is a guaranteed MCP in yellow which clears the power market in the APN.

To calculate the aggregated power supply and consumption, the cumulative PPF for an APN in Definition 4.3 is introduced.

Definition 4.2: Price-to-Power Function [SGH24]

The price-to-power function (PPF) is a mathematical function which represents the offers O_i and bids B_j and holds

$$PPF : p \rightarrow P \text{ with } p \in [p_{min}, p_{max}] \text{ and } P \in \mathbb{R}. \tag{4.8}$$

Accordingly, a PPF describes the offering behavior O_i of a seller i or the bidding behavior B_j of a bidder. The PPF is compliant with the three rules R1-R3 defined in (4.4), (4.5), and (4.6).

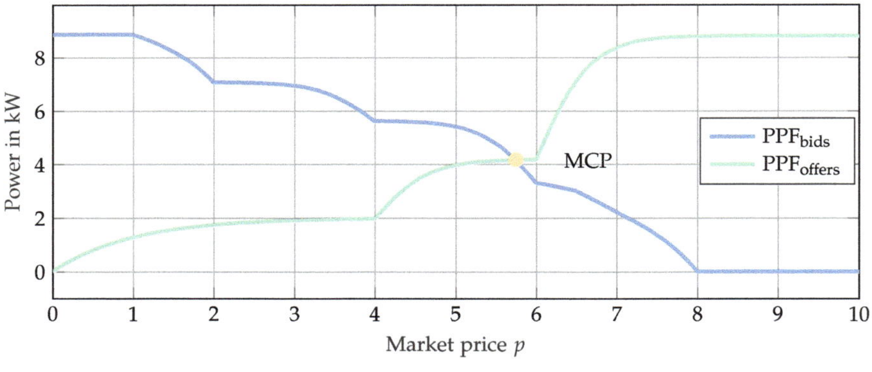

Figure 4.4: Aggregated PPFs for the bids and the offers in the defined price range $p \in [0, 10]$ and the resulting MCP highlighted with the yellow dot. The bids and offers are in accordance to the proposed rules R1-R3 in (4.4), (4.5), and (4.6).

Definition 4.3: Cumulative Price-to-Power Function [SGH24]
The cumulative price-to-power function (PPF) of an APN is given by

$$PPF_{cum} = PPF_{offers} - PPF_{bids} = \sum_{i \in \mathcal{S}} PPF_i - \sum_{j \in \mathcal{B}} PPF_j \tag{4.9}$$

$$\text{with } APN = \{\mathcal{S}, \mathcal{B}\}.$$

The PPF_{cum} in Figure 4.5 represents the aggregated power supply and consumption over the valid market price range $p \in [0, 10]$. Moreover, the MCP and the according electric power $\text{PPF}_{\text{cum}} = 0$ is highlighted with the yellow dashed lines. By the reformulation, the MCP is no longer the intersection of $\text{PPF}_{\text{offers}}$ and PPF_{bids}, but the intersection of the PPF_{cum} with $P = 0$. Consequently, PPF_{cum} describes the power balance in the APN in accordance with the market price p. As formulated in Definition 4.4, the MCP indicates the price p for which $\text{PPF}_{\text{cum}} = 0$ holds.

Definition 4.4: Market Clearing Price Determination [SGH24]
Based on the PPF_{cum} in Definition 4.3, the market clearing price determination is expressed by

$$MCP = \underset{p}{\text{argmin}} \, |PPF_{cum}(p)|. \tag{4.10}$$

Price-to-Power Function Design Rules

The PPFs as formalized in Definition 4.2 represent the relation between the market price and the power consumption or supply of a component in the APN. In the auction mechanism, the PPFs facilitate the individual behavior of each component within the provided framework. From a global perspective, the PPF for a seller $\mathcal{S}$ offering power is monotonically increasing as depicted in Figure 4.4 by the green line. On the other hand, the PPF for a bidder $\mathcal{B}$ consuming power is defined as a monotonically decreasing function as visualized with the blue line.

To ensure a reliable power supply of the safety-relevant components in the APN, price ranges for the individual PPF design are given in Table 4.1. These mandatory design rules complement the predefined rules concerning the general UPA in Definition 4.1. Table 4.1 lists the specified price intervals used in this thesis and the abstract interpretations regarding the component and APN behavior. The proposed price intervals and the respective limits are specified for the valid price interval $p \in [0, 10]$ and visualized in Figure 4.5.

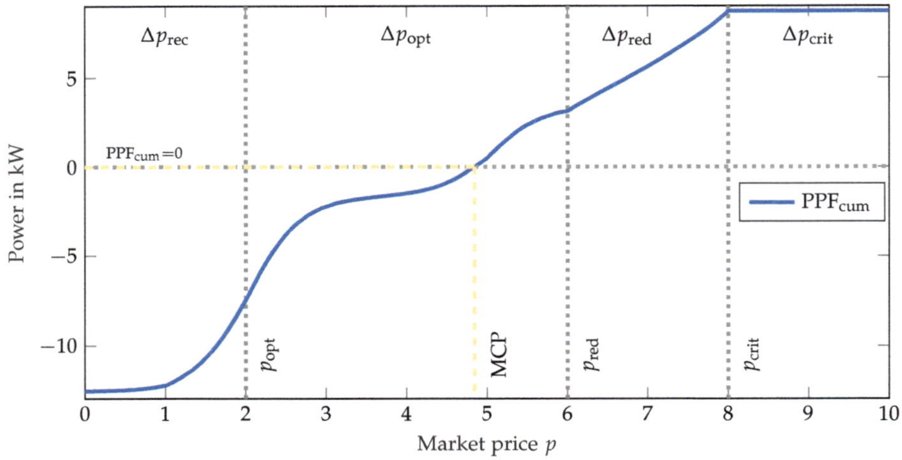

Figure 4.5: Cumulative PPF as the aggregation of offers and bids for an exemplary APN with the MCP for $\text{PPF}_{\text{cum}} = 0$ and the price intervals Δp_{rec}, Δp_{norm}, Δp_{red}, and Δp_{crit} defined in Table 4.1.

The price range Δp_{rec} indicates that the power suppliers in the APN offer a high amount of power. This may be the case when the EM of an HEV is or has recently been in recuperation mode. To convert as much kinetic energy as possible, the flexible systems and comfort components are asked to utilize the available power. As a result, more energy is recovered, or the battery charging current is reduced, mitigating the electro-chemical strain on the battery or other storages [MOG+09, VA15, RSSH21].

The price interval Δp_{norm} depicts the normal operation mode and desired APN status. In this price range, the various components are expected to be in their normal

Table 4.1: Price partitioning for the proposed market price range $p \in [0, 10]$ and the correlated component or system behavior as rules for individual PPFs in a reliable auction framework.

Range	Variable	Interpretation
$[0, 2]$	Δp_{rec}	Recuperation mode with extra power available and the request for increased demand.
$(2, 6]$	Δp_{norm}	Adaptive power supply and normal operation for the components and the power supply.
$(6, 8]$	Δp_{red}	Reduced comfort and reduction of demand in order to stabilize the power supply.
$(8, 10]$	Δp_{crit}	Critical APN status and further reduction of demand to ensure safe driving operation in an emergency.

operation points. At the same time, the power supply is commonly covered by the EM in generator mode and the energy storages. For a rising power demand in the APN or if the power supplying systems reach their limits, the price level exceeds Δp_{norm} and enters Δp_{red}. The power-supplying components are no longer able to provide enough power for the overall demand. Therefore, the components providing comfort functionalities reduce their power demand with regard to an increasing market price and their respective PPF, so the power in the APN is mainly balanced by demand reduction. Nevertheless, the additional provision of power from the storage systems may be another measure to balance the power. At the same time, this option puts more strain on the storage systems, leading to faster aging [MOG+09].

The last price range Δp_{crit} highlights a critical APN state, in which the vehicle is in emergency mode. In this work, emergency mode describes a vehicle state where safety-relevant components may fail due to insufficient power supply or a critical voltage drop. Power management is particularly relevant during these phases to ensure a proper power supply. A design rule for the PPFs is that if the critical price level Δp_{crit} is reached, every comfort load in the APN has to switch off. The APN operates in emergency mode and, thus, only safety-relevant systems are supplied with power, ensuring a safe driving operation until the vehicle standstill. The price limits p_{norm}, p_{red}, and p_{crit}, as highlighted in Figure 4.5, refer to the lower limit of the according price intervals Δp_{norm}, Δp_{red}, and Δp_{crit}.

The mandatory design rules for the individual PPFs form a reliable framework for the power balancing in the APN. Hence, the reliable behavior of all components and systems is guaranteed, assuring a robust power supply in every situation. With this reliable framework, the proposed auction-based power management provides the desired plug-and-play integration property since new components are seamlessly integrated into the group of sellers $\mathcal{S}$ or bidders $\mathcal{B}$. Thus, the new components are another element in the sums of the cumulative PPF in (4.9). On the other hand, the mandatory behavior in the price ranges Δp_{red} and Δp_{crit} enhance the robust power supply in critical situations independent of the specific APN configuration.

4.1.2 Extension for Multiple Voltage Levels and Overall Mechanism

The proposed auction mechanism with the UPA in Definition 4.1 and the MCP determination in Definition 4.4 achieve the power balance for a single voltage level which comprises all components in the APN. The mandatory design rules also ensure a reliable power balancing framework to enhance the power supply for safety-relevant components during power shortages.

In the case of multiple voltage levels, there is the possibility and need to transfer power between the power levels via DC/DC-converters. Thus, in the HEV configuration in Figure 3.2, there is only one electric machine (EM) in the 48V level, which supplies both voltage levels. However, the power transfer via DC/DC-converters is limited, and the limitations vary in terms of thermal stress, wearing effects, or acute failures [AK14, THVY16]. Therefore, the DC/DC-converter limitations $P_{\mathrm{PE,\,lim}}$ have to be taken into account in the power management of vehicles with multiple voltage levels.

Therefore, the auction mechanism is adapted for multiple voltage levels

$$\mathbb{V} = \{1, 2, \ldots, v\}. \tag{4.11}$$

The power transfer between two voltage levels l and m is described by

$$P^{lm}(k) \text{ with } l, m \in \mathbb{V} \text{ and } l \neq m. \tag{4.12}$$

Here, the first index l determines the providing voltage level, and the second index m represents the receiving voltage level. Since most DC/DC-converters between voltage levels are bidirectional, the power transfer is possible in both directions P^{lm} and P^{ml} [ELR08]. First, the auction procedure for multiple voltage levels determines the MCP for the whole APN. If a given power transfer limit $P^{lm}_{\mathrm{PE,\,lim}}$ is exceeded within the achieved global power balance, further market price adjustments are necessary. Thus, an individual market price p^m is determined for each voltage level $m \in \mathbb{V}$.

To determine the power transfer P^{lm}, the cumulative PPF in Definition 4.3 is applied to the individual voltage levels:

$$\mathrm{PPF}_{\mathrm{cum},\,m}(p) = P^{lm} \text{ with } p = \mathrm{MCP} \text{ with } m, l \in \mathbb{V} \text{ and } l \neq m. \tag{4.13}$$

By the individual cumulative PPF in (4.13), which is evaluated for the global MCP, a power shortage for $P^{lm} < 0$ or a power surplus[8] for $P^{lm} > 0$ in the voltage level m is calculated. Based on the APN architecture, the power difference P^{lm} is transferred via the DC/DC-converter from voltage level l toward the supplied level m. Hence, the voltage level l has to provide a power surplus

$$\mathrm{PPF}_{\mathrm{cum},\,l}(p) = -P^{lm}. \tag{4.14}$$

In order to meet the power transfer limitations $P^{lm}_{\mathrm{PE,\,lim}}$ in the APN, the adaptation procedure starts with the last supplied voltage level $m \in \mathbb{V}$ in the APN architecture. Subsequently, the connected voltage levels starting with the supplying level l are adapted. Similar to the procedure for the global MCP determination in Definition 4.4, the cumulative $\mathrm{PPF}_{\mathrm{cum},\,m}$ is built for the individual power level m with its incorporated components. To determine a suitable market price p^m for the individual voltage level m, the prevailing power transfer limit $P^{lm}_{\mathrm{PE,lim}}$ is considered:

$$p^m = \underset{p}{\operatorname{argmin}} \, |\mathrm{PPF}_{\mathrm{cum},\,m}(p) + P^{lm}_{\mathrm{PE,\,lim}}|. \tag{4.15}$$

[8] In case of a power surplus in voltage level m, the power transfer is directed toward voltage level l which is expressed by the switched order of the indices l and m leading to P^{ml}.

With the new market price p^m, the power transfer limitation is met by

$$\text{PPF}_{\text{cum},\,m}(p^m) = -P^{lm}_{\text{PE, lim}} = -P^{lm}. \tag{4.16}$$

In the process, the increasing market price p^m affects a reduction in power consumption and a stronger exploitation of the supplying components in the voltage level m. If the power transfer limitation $P^{lm}_{\text{PE, lim}}$ is met, the supplying voltage level l is adjusted accordingly. Since the power P^{lm} to supply the connected voltage level m decreased, there is the possibility and need for a new UPA to determine an individual market price p^l:

$$p^l = \underset{p}{\text{argmin}}\, |\text{PPF}_{\text{cum},\,l}(p) - P^{lm}_{\text{PE, lim}}|. \tag{4.17}$$

This procedure is continued until all power transfer limitations $P^{lm}_{\text{PE, lim}}$ for $l, m \in \mathbb{V}$ are met.

Overall Procedure

To calculate the global MCP in (4.10), an iterative algorithm for the market price adaptation with the step size γ is used. The procedure in Definition 4.5 applies similarly for the individual market prices in (4.15) and (4.17).

Definition 4.5: Market Clearing Price Determination: An Iterative Algorithm [SGH24]

For the market clearing price (MCP) determination within an iterative algorithm, the formulation in (4.10) is approximated by

$$|PPF_{cum}(p)| < \epsilon \text{ for } p = MCP. \tag{4.18}$$

For the market price adaptation during the iterative procedure, the following rules apply:

1. *If* $PPF_{cum}(p) < -\epsilon$, *set* $\tilde{p} = p + \gamma$,
2. *If* $PPF_{cum}(p) > \epsilon$, *set* $\tilde{p} = p - \gamma$,

with $\gamma \in \mathbb{R}^+$, $\epsilon \in \mathbb{R}^+$, *and* $\forall p, \tilde{p} \in [0, 10]$. *These rules are repeatedly applied until the condition in (4.18) holds.*

The definition of the step size γ and the parameter ϵ is a trade-off between the power balance accuracy and the algorithm speed. For a guaranteed convergence, γ and ϵ have to fulfill the condition:

$$|\text{PPF}_{\text{cum}}(p) - \text{PPF}_{\text{cum}}(p+\gamma)| < 2\epsilon, \gamma \in \mathbb{R}^+, \epsilon \in \mathbb{R}^+, \text{ and } \forall p \in [0, 10] \tag{4.19}$$

as depicted in [SGH24]. This condition ensures that the region defined by ϵ is reach during the iterative procedure. In this regard, the condition takes into account the step size γ and the derivative $\partial \mathrm{PPF}_{\mathrm{cum}}(p)/\partial p$ expressed by the absolute value of $|\mathrm{PPF}_{\mathrm{cum}}(p) - \mathrm{PPF}_{\mathrm{cum}}(p+\gamma)|$ [SGH24].

Figure 4.6 gives an overview of the proposed auction mechanism for APNs with multiple voltage levels. The overall procedure includes the initialization of the UPA, the iterative execution to determine the global MCP, if necessary, the determination of individual MCPs for the voltage levels, the communication to the participants, and their adaptation according to the MCP and the PPFs. To continuously coordinate the power consumption and supply in the APN, the auction procedure is periodically repeated with a cycle time of $\Delta T = 200\,\mathrm{ms}$ as proposed in [Gra04a]. Consequently, the cycle time ΔT is a parameter that adjusts the power management procedure. It is a trade-off between a higher calculation and communication frequency, on the one hand, and a more precise power balancing due to a smaller sample time, on the other hand. The power consumption and supply is regarded constant within the intervals $k \in \mathbb{N}^+$ of length ΔT. As a result, all disturbances within an interval k cause certain deviations from the calculated power balance $\mathrm{PPF}_{\mathrm{cum}}(p_k) = 0$.

During the initialization, the individual PPFs are updated based on the current information, and the initial price is set to $p = p_{\mathrm{init}}$. To examine the next MCP, the $\mathrm{PPF}_{\mathrm{cum}}$ is calculated. According to Definition 4.5, the market price for the next time interval p_k is adjusted until supply and demand are equal for the overall power network, which is defined as $|\mathrm{PPF}_{\mathrm{cum}}| < \epsilon$. With the UPA Definition 4.1, the power balance is guaranteed within the valid market price range $p \in [0, 10]$.

For vehicles with multiple voltage levels, the power transfer limitations $P_{\mathrm{PE,lim}}$ of the connecting DC/DC-converters have to be taken into account. If these constraints are not met, the market price has to be further adjusted for the individual voltage levels. Consequently, the algorithm provides individual market prices p_k^m and p_k^l for the connected voltage levels $m, l \in \mathbb{V}$ until all power transfer limitations $P_{\mathrm{PE,lim}}^{lm}$ are met. Finally, the global MCP or the individual MCPs for the voltage levels are communicated to the components in the APN. Regarding the MCP and the individual PPFs, all components $i \in \mathrm{APN}$ adapt their power supply and consumption for the next time interval k. The last market price p_{k-1} serves as the initial price for the next iteration, leading to a faster convergence since the deviations between two successive intervals k are often negligible.

4.1.3 Individualization by Price-to-Power Functions

Despite the centralized determination of the MCP in Figure 4.6, individual and distributed decisions are key to facilitate adaptations to user and passenger preferences [SRSH21, SSH21]. These decisions are enabled by the design of individualized PPFs for the participants in the power auction. For overall stability, every PPF follows the

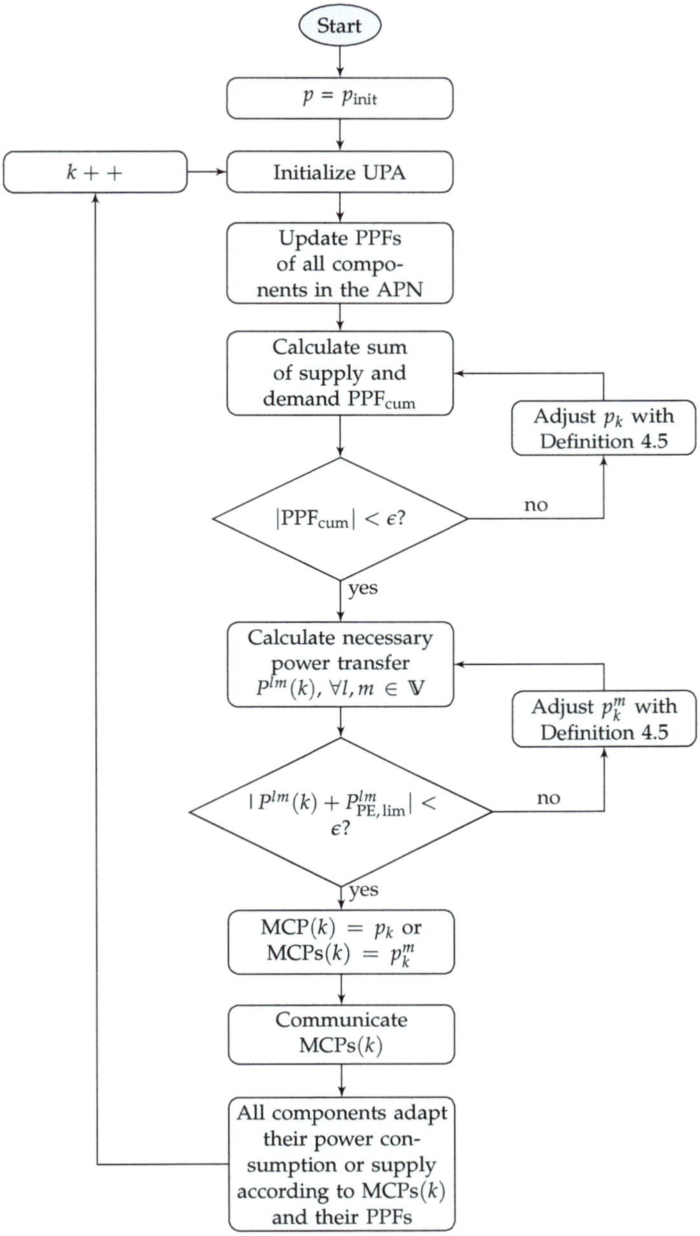

Figure 4.6: Proposed automotive power management based on the UPA with the adaptation for multiple voltage levels $\mathbb{V}_j$ and the consideration of PE limitations $P_{\text{PE, lim}}$.

mentioned rules and design patterns marked by the price ranges in Table 4.1. Within these constraints, every component's PPF may vary due to optimized operation or user preferences.

The component function is an important factor concerning the PPF. As depicted in Figure 4.1, there is a fundamental division between sellers and bidders providing or consuming electric power, respectively. However, the distinction between sellers and bidders is derived from the UPA theory [Fri60]. Since the storage systems consume or provide power depending on the situation, they may be part of both groups and are commonly assigned to the sellers. Additionally, the bidder components belong to different load categories, which are visualized in Figure 3.8.

So, the bidders are subdivided into comfort or hotel loads and safety-relevant loads. The flexible loads represent a subgroup of the comfort loads and are able to shift power consumption to some extent. In general, the sellers have a positive PPF during the provision of power, while the bidders have a positive PPF displaying the power consumption over the price. This has to be taken into account in (4.10) when determining the sum of power supply and consumption and deriving the current MCP.

For the individual PPF, the inherent system states, the user preferences, and the current user inputs are considered. The general user preferences are also relevant for the hysteresis mechanism, which is elaborated in the following section. Nevertheless, the price ranges in Table 4.1 prescribe the mandatory specifications for every PPF, ensuring a reliable power management framework in all situations. In the following, the PPFs of exemplary sellers and various bidders are mathematically defined and visualized for the mutual price range $p \in [0, 10]$.

Electric Machine

The EM is the leading supplier of electric power in the APN of an HEV. Since this work focuses on the electric architecture of an HEV, the PPF for the EM plays a critical role. Besides the power supply in generator mode, the EM offers opportunities for boosting or recuperation depending on the kind of HEV. In boost mode, the EM is deployed to produce extra torque and support the driving operation. As a result, the EM changes its role from the primary seller to the primary bidder in the power auction, represented by a strongly varying and switching PPF. For recuperation or regenerative braking, the HEV driving control utilizes the EM to put negative torque on the mechanical powertrain and, thus, decelerate the vehicle. In this process, the EM generates additional electric power, converting mechanical energy to electrical energy, which is then available in the power network. The actual amount of electric power correlates with the negative torque and thereby depends on the torque distribution between the different braking systems [EGLE18]. To increase the driving efficiency, the ratio for regenerative braking should be as high as possible [HWK+17].

With regard to the different modes, the EM has three distinct PPFs, which depict the respective behavior of the EM. In Figure 4.7, exemplary PPFs for the three modes are visualized. While operating as a seller in generator mode, the EM adapts its power supply based on the required power shown by the blue graph. As mentioned before, the PPF for generator mode is positive since the EM is regarded as a seller that offers electric power in the proposed auction. In this mode for adaptive power supply, the PPF for the market price p is given by the exponential growth function

$$P_{\text{ada}}(p) = P_{\text{gen}} \cdot (1 - e^{-p\lambda_{\text{gen}}}). \tag{4.20}$$

Accordingly, the adaptive supply $P_{\text{ada}}(p)$ grows smoothly with the required power in the APN, which is expressed by the current market price p. The growth factor λ_{gen} adjusts the slope. Since the generation of electric power results in a negative torque, the maximal power P_{gen} for the EM in generator mode is subject to the control algorithm for the powertrain [RL07, GS17]. The choice of P_{gen} includes the EM states, for example, the motor speed ω, or other driving strategy parameters as the allowable deceleration through negative torque M_{gen} [EGLE18].

In boost mode, the EM deploys additional torque to the powertrain, supporting the driving operation. The function of this mode

$$P_{\text{boost}}(p) := \begin{cases} P_{\text{set, boost}}, & \text{for } p \in \Delta p_{\text{rec}} \vee \Delta p_{\text{norm}}, \\ P_{\text{set, boost}} \cdot (1 - \frac{p - p_{\text{red}}}{p_{\text{crit}} - p_{\text{red}}}), & \text{for } p \in \Delta p_{\text{red}}, \\ 0, & \text{for } p \in \Delta p_{\text{crit}}, \end{cases} \tag{4.21}$$

describes the high electric power demand with reference to the aimed acceleration. The actual power demand P_{set} is set by the powertrain control in coordination with the energy management. In case the energy storages are not able to provide the required

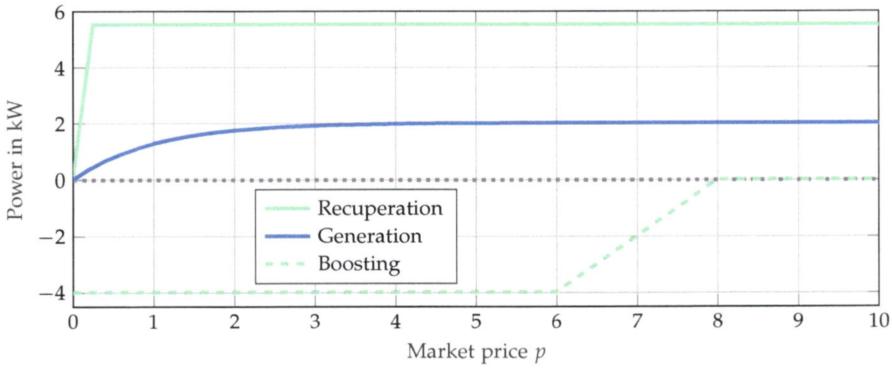

Figure 4.7: PPF of the EM with the three modes recuperation (rec), generation (ada), and boosting (boost) and exemplary graphs for P_{rec}, P_{ada}, and P_{boost}.

power and the normal price range Δp_{norm} is exceeded, the power consumption in the price interval Δp_{red} is reduced until a power balance is achieved. Boosting is prohibited during a critical phase represented by Δp_{crit}.

In recuperation mode, the EM supports the vehicle's deceleration through negative torque. Simultaneously, the EM recovers mechanical power and generates electric power in this mode. Consequently, the PPF for regenerative braking holds

$$P_{\text{rec}}(p) := \begin{cases} P_{\text{set, rec}} \cdot \frac{p}{p_{\text{crit}}/4}, & \text{for } p \in [0, p_{\text{crit}}/4], \\ P_{\text{set, rec}}, & \text{for } p \in [p_{\text{crit}}/4, 10], \end{cases} \tag{4.22}$$

and sets the recuperation power P_{rec}, which is constant for all market prices apart from the interval $p \in [0, p_{\text{crit}}/4]$. The resulting electric power P_{rec} is depending on the braking system. If the amount of regenerated power P_{rec} exceeds the storage potential and the aggregated consumption (in the price range $p \in [0, p_{\text{crit}}/4]$), the braking torque distribution between the EM and the hydraulic brake, for instance, is adjusted by the powertrain control [EGLE18]. As a result, the deceleration momentum distribution is shifted to other braking systems [EGLE18].

In summary, the EM behavior mainly depends on the three motor modes. Thus, the three PPFs for boosting (boost) in (4.21), for adaptive power generation (ada) in (4.20), and for recuperation (rec) in (4.22) determine the role of the EM in the power auction. Therefore, the combined PPF is

$$\text{PPF}_{\text{EM}}(p) := \begin{cases} P_{\text{boost}}(p), & \text{in boost mode,} \\ P_{\text{ada}}(p), & \text{for adapted power supply,} \\ P_{\text{rec}}(p), & \text{in recuperation mode.} \end{cases}$$

Battery Storages

Apart from the EM, the battery storages are commonly the main energy source in the APN [Nea20]. By design, battery storage as a passive electrochemical system depends on the prevailing voltage. The battery is charged if the voltage is higher than the open circuit voltage (OCV). On the other hand, the battery is discharged for voltages beneath this voltage level and provides power to the power network. Thus, the battery storage acts as a seller or bidder in the power auction. For further details regarding the physical behavior, it is referred to Section 3.2.2.

The battery behavior depends on the internal states SOC, SOH, and the temperature T_{Batt}. Therefore, the PPF

$$\text{PPF}_{\text{Batt}}(p) = f(\text{SOC}, \text{SOH}, \lambda_{\text{Batt}}(T_{\text{Batt}}), p),$$

which is visualized in Figure 4.8, comprises these internal states of the battery. According to the current market price p in the UPA, the battery behavior smoothly shifts

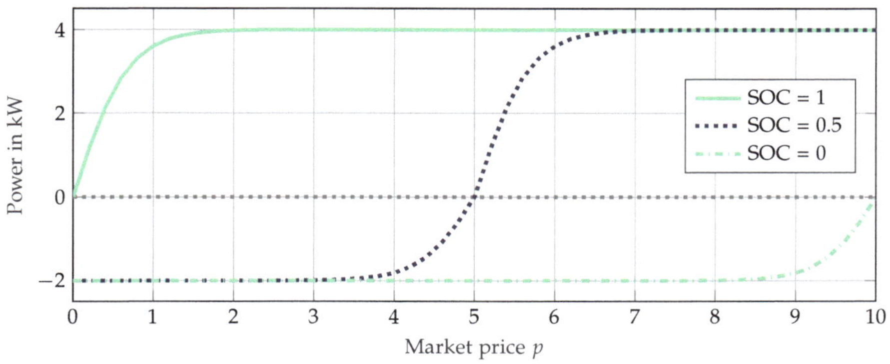

Figure 4.8: Individual PPF for the battery with different SOC levels. Besides the SOC, the battery PPF depends on the SOH and the battery temperature T_{Batt}.

from charging to discharging with an increasing power demand in the APN. Hence, the battery storage mainly stores power in phases with sufficient power supply and emits power during phases with power deficiency.

To account for different temperatures T_{Batt} and their effect on the battery capacity C_{Batt}, the factor λ_{Batt} is specified. For the implementation of the PPF, a sigmoid function with

$$f(x) = \frac{1}{1+e^{-x}}$$

is chosen, which suitably describes a smooth transition from the negative zone to the positive zone. The customized sigmoid function of an exemplary lithium-ion battery pack is given by

$$\text{PPF}_{\text{Batt}}(p) := \begin{cases} P_{\text{c}} \cdot (1 - \frac{2}{1+\exp(\lambda_s(-p+p_{\text{x}}))}), & \text{for } p \leq p_{\text{x}}, \\ P_{\text{dis}} \cdot (\frac{2}{1+\exp(\lambda_s(-p+p_{\text{x}}))} - 1), & \text{for } p_{\text{x}} < p. \end{cases} \tag{4.23}$$

The limited charging power P_{c} and discharging power P_{dis} depend on the SOH and the temperature T_{Batt}. With a decreasing SOH, the charging and discharging currents are alleviated to reduce physical strain and aging [MOG+09, RSSH21]. To describe the charging and discharging behavior independently, the PPF is divided into two parts. The price p_{x} marks the transition between consumption and supply in the battery behavior and is derived from the current SOC. The PPF gradient is adjustable by λ_{s}, which shapes the transient behavior during the auction procedure. Exemplarily, the parameter is chosen to $\lambda_s = 3$, resulting in the graphs in Figure 4.8.

The limit for the charging power is expressed by

$$P_{\text{c}} = P_{\text{c,max}} \cdot \lambda_{\text{Batt,c}}(T_{\text{Batt}}) \cdot \text{SOH},$$

with SOH $\in [0,1]$. The maximum charging power $P_{c,\max}$ is an individual variable, depending on the cell characteristics and the design of the battery pack. The temperature factor of charging

$$\lambda_{\mathrm{Batt,c}} := \begin{cases} 0, & \text{for } T_{\mathrm{Batt}} < T_{\mathrm{c,min}} \\ 0.95 + 0.025\, T_{\mathrm{Batt}}, & \text{for } T_{\mathrm{c,min}} \leq T_{\mathrm{Batt}} \leq T_{\mathrm{c,max}}, \\ 0, & \text{for } T_{\mathrm{c,max}} < T_{\mathrm{Batt}} \end{cases}$$

describes the drop in the battery voltage with a decreasing temperature T_{Batt} [GLD02]. Furthermore, the minimum and maximum temperature regarding the safe charge operation $T_{\mathrm{c,min}}$ and $T_{\mathrm{c,max}}$ have to be taken into account. Otherwise, wearing effects, leading to accelerated aging of the battery pack, may occur [GLD02, MOG+09]. According to the discharging, the factor for the temperature is given by

$$\lambda_{\mathrm{Batt,dis}} := \begin{cases} 0, & \text{for } T_{\mathrm{Batt}} < T_{\mathrm{dis,min}} \\ 0.95 + 0.025\, T_{\mathrm{Batt}}, & \text{for } T_{\mathrm{dis,min}} \leq T_{\mathrm{Batt}} \leq T_{\mathrm{dis,max}}, \\ 0, & \text{for } T_{\mathrm{dis,max}} < T_{\mathrm{Batt}}. \end{cases}$$

With $\lambda_{\mathrm{Batt,dis}} = 0$ for temperatures outside the safe temperature area, the battery storage is not allowed to provide or consume power. The minimum and maximum temperature for the discharging process are given by $T_{\mathrm{dis,min}}$ and $T_{\mathrm{dis,max}}$, respectively.

Similar to the charging power P_{c}, the discharging power is restricted by

$$P_{\mathrm{dis}} = P_{\mathrm{dis,max}} \cdot \lambda_{\mathrm{Batt,dis}}(T_{\mathrm{Batt}}) \cdot \mathrm{SOH},$$

paying attention to the temperature, the SOH, and the maximum discharging power $P_{\mathrm{dis,max}}$ of a specific battery pack. In general, for lithium-ion cells, the rated discharging power P_{dis} is higher than the rated charging power P_{c} [Nea20].

The price p_{x} which separates charging and discharging operation is derived from the SOC by

$$p_{\mathrm{x}} = \lambda_x \cdot (1 - \mathrm{SOC})$$

for a prescribed price range $p \in [p_{\min}, p_{\max}]$ and the SOC $\in [0,1]$. The factor λ_x is derived from the price range by $\lambda_x = p_{\max} - p_{\min}$ which results to $\lambda_x = 10$ for the set price range $p \in [0,10]$. With the parameter design of p_x, the battery PPF accounts for the necessity of charging when the current SOC is low. Concerning the extreme values of the SOC, the battery is only a power supplier (SOC $= 1$) or a power consumer (SOC $= 0$) as highlighted in Figure 4.8 by the green and the green dashed graph, respectively.

Heating, Ventilation and Air Conditioning

According to Section 3.2.4, the HVAC system is mainly a comfort load and one of the major power consumers in terms of peak power and, in particular, average power consumption [VA15]. An advantageous property is that it offers some flexibility due to the fact that humans are not able to precisely sense temperature and rarely notice little temperature deviations in the range of 1°C [LLDL11, RBW+12, VA15]. Further, the temperature behavior of the HVAC system comprises large time constants resulting in the possibility to shift heating or cooling operation in a particular time window [VA15]. These time constants result from the inherent energy storage since the cooling cycle and the vehicle cabin act like temperature capacities. When shifting power demand, the HVAC system uses its thermal buffers to store electrical power by increasing the cooling process if power is available and reducing the cooling process if power is restricted. Consequently, the HVAC PPF in Figure 4.9 depicts power adaptations in the price ranges Δp_{rec} and Δp_{red}. Therefore, the HVAC system offers the desired flexibility in the proposed auction mechanism for power balancing.

The PPF of the HVAC system is expressed by

$$\mathrm{PPF}_{\mathrm{HVAC}}(p) := \begin{cases} P_{\max}, & \text{for } 0 \leq p \leq p_{\text{norm}}/2, \\ P_1(p), & \text{for } p_{\text{norm}}/2 < p \leq p_{\text{norm}}, \\ P_{\text{norm}}, & \text{for } p \in \Delta p_{\text{norm}}, \\ P_2(p), & \text{for } p \in \Delta p_{\text{red}}, \\ 0, & \text{for } p \in \Delta p_{\text{crit}}. \end{cases} \tag{4.24}$$

The market price p_{norm} describes the border between Δp_{rec} and Δp_{norm}. The intervals of P_1 and P_2 in (4.24) are critical design parameters regarding the HVAC behavior within the auction mechanism. A gradual transition between the three power levels

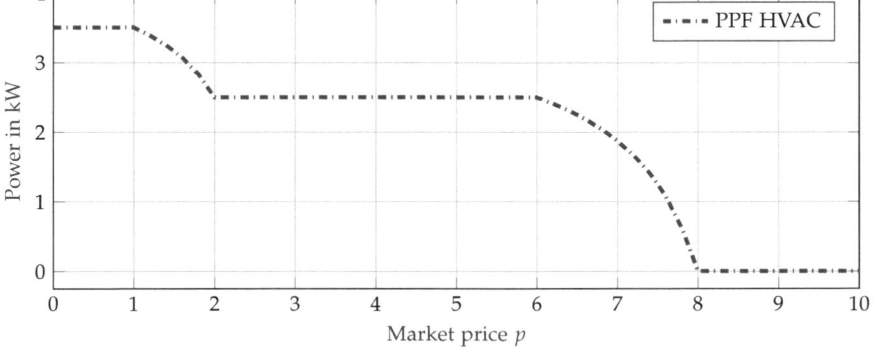

Figure 4.9: Customization of the HVAC PPF with power adaptations in the price ranges Δp_{rec} and Δp_{red}.

$P_{\max}$, P_{norm}, and zero is ensured by

$$P_1(p) = P_{\max} + (P_{\max} - P_{\text{norm}}) \cdot \left(\frac{p - p_{\text{norm}}/2}{p_{\text{norm}}/2} \frac{1}{p - p_{\text{norm}} - 1} \right)$$

and

$$P_2(p) = P_{\text{norm}} \cdot \left(1 + \frac{p - p_{\text{red}}}{p_{\text{crit}} - p_{\text{red}}} \frac{1}{p - p_{\text{crit}} - 1} \right).$$

These gradual adaptations of the power consumption $P_1(p)$ and $P_2(p)$ are located on the left and the right side of the price interval $\Delta p_{\text{norm}} \in (2, 6]$, which marks the preferred operation area with the normal power consumption P_{norm}. The power P_{norm} for normal operation depends on user preferences, user inputs, and the HVAC control strategy [KLFE11]. Below the price $p = 2$ (in Δp_{rec}), the supply exceeds the preset consumption for normal operation. Hence, the HVAC system as a flexible load in the APN provides the possibility to consume additional power as long as the user requirements are continuously met. The flexible consumption is constrained by the maximal power of the HVAC system $P_{\max}$.

In case the supplying systems are at their capacity boundaries indicated by a market price $p > 6$ (in Δp_{red}), the power consumption decreases smoothly with an increasing market price. Thus, the HVAC system reacts to a presumable power limitation in the APN to support the power balancing.

The required parameters or states are the ambient temperature T_{amb}, the cabin temperature T_{cab}, the preferred temperature T_{set}, and the maximal power consumption of the HVAC system $P_{\text{HVAC,max}}$. These parameters are considered to calculate the normal power demand P_{norm} and the maximum power $P_{\max}$. The actual power demand P_{norm} mainly depends on the desired cabin temperature T_{cab} and the surrounding circumstances, such as the ambient temperature T_{amb} or the humidity.

Seat Heating

While the HVAC system gradually varies its power consumption, the seat heating system is a bidder with discrete power steps due to the switching between different heating resistors [Rei11, Nea20]. Hence, the integration of the seat heating in the auction mechanism in Figure 4.6 causes challenges since the MCP determination in (4.10) requires continuous PPFs. With stepwise switching, the precise calculation of a power balance is challenging or, under certain circumstances, impossible [Büc08]. To facilitate a smooth market price calculation, the seat heating PPF visualized in Figure 4.10 comprises linear transitions between the discrete power levels.

The depicted seat heating system consists of three power levels, which are denoted by $P_{\text{Seat},1}$, $P_{\text{Seat},2}$, and $P_{\text{Seat},3}$ with increasing power demand. If the seat heating is turned off, there is no standby current, resulting in $P_{\text{Seat},0} = 0$. In Figure 4.10, the second

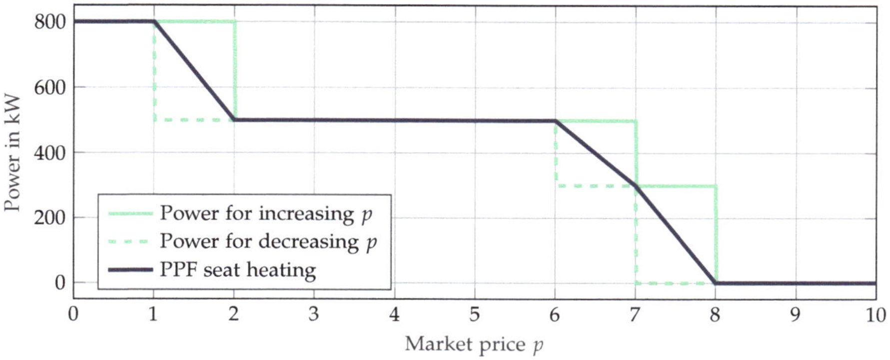

Figure 4.10: Customized PPF for the seat heating with the representative PPF in dark blue and the diverging power levels with hysteresis in green for an increasing market price p (solid line) and a decreasing price p (dashed line).

power level $P_{\text{Seat},2} = 500\text{W}$ is applied in the normal price range Δp_{norm}. The applied power level in Δp_{norm} is determined by the passenger inputs. For additional power consumption in a recuperation phase, the highest seat heating level $P_{\text{Seat},3} = 800\text{W}$ is available. On the other side, if the market price p exceeds the threshold p_{red}, the heating level is reduced to $P_{\text{Seat},1}$ or $P_{\text{Seat},0}$ with respect to the price p.

To prevent cyclic switching, a hysteresis is implemented for the discrete power levels highlighted with the green lines in Figure 4.10. Thus, the power level is kept for an increasing market price until the next price threshold is reached. For a decreasing market price, the heating level is switched directly. As a result, in the case of limited power availability, slight adaptations are first made by the continuously controlled loads, such as the HVAC system. Hence, the switching loads are not directly switched to a lower level or turned off. The exact placement of the hysteresis intervals within the market price range $p \in [p_{\text{min}}, p_{\text{max}}]$, as well as the hysteresis range, are design parameters in the auction framework.

Blowers

Like the described seat heating system, the blower PPF in Figure 4.11 may be implemented as a switching component with several discrete power levels [Rei11, Nea20]. Since the blowing system commonly needs a more distinguished adjustment to meet the user preferences, it includes more discrete power levels. In contrast to the seat heating, variations in the blower power are easier to notice. Therefore, additional power consumption within the auction mechanism is not intended for the blowers. Furthermore, the degradation of the blowers is delayed as long as possible due to the direct effect on passenger comfort and awareness. So, the order of switching off has

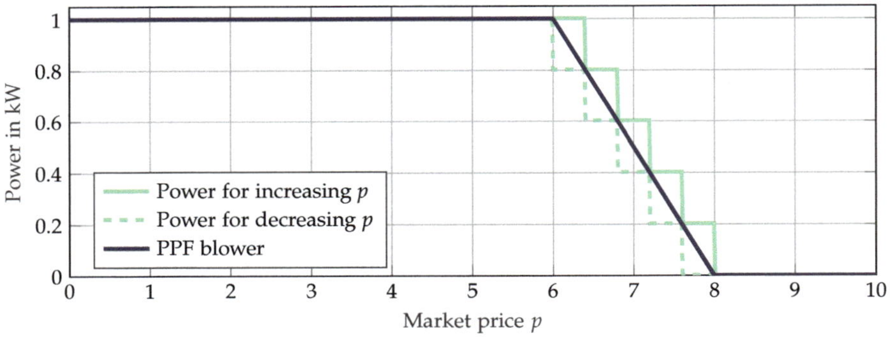

Figure 4.11: Individual PPF of the blower in dark blue with switching behavior in the price interval $\Delta p_{\text{red}} \in (6,8]$. The hysteresis to prevent cyclic switching is depicted in green for an increasing price p (solid line) and a decreasing price p (dashed line).

to be taken into account in the overall auction mechanism to ensure user comfort and acceptance.

Like the seat heating PPF, the blower PPF comprises linear transitions between the power stages and a hysteresis design to inhibit oscillation between neighboring stages. Again, the current blower level is held until the next threshold is reached. Due to the simpler user awareness, this arrangement is even more critical for the blower system than for the seat heating. The visualization in Figure 4.11 depicts the general PPF design with a blower level $P = 1\text{kW}$, which is subject to the HVAC automatic or the passenger inputs. Hence, the actual blower PPF changes with reference to the settings chosen by the users.

Safety-relevant and Entertainment Systems

Except for the comfort loads, the APN comprises safety-relevant components and systems as described in Section 3.2.4. These systems are required for a safe driving operation so their power demand is non-negotiable in the proposed auction for electric power.

Therefore, the associated PPFs for safety-relevant functions is given by a constant bid for power over the whole price range $p \in [0, 10]$. On the other hand, if the price level exceeds the price p_{red}, the comfort components reduce their power demand, ensuring a stable voltage level and the proper functioning of the safety-relevant systems. Hence, the power supply of the highly relevant systems for safe driving operation is guaranteed by the auction mechanism, the PPF design, and the price ranges for mandatory PPF patterns in Table 4.1.

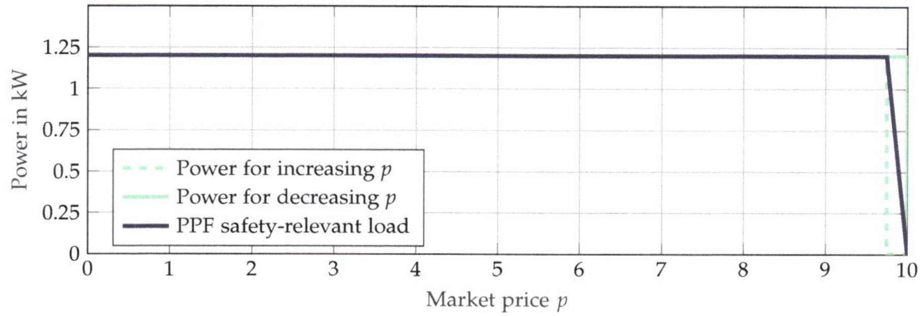

Figure 4.12: Particular PPF design for a safety-relevant load that consumes a non-negotiable amount of electric power and is independent of the market price p.

Certain power levels depend on the operating and control strategies and are not negotiable within the power management. A PPF for a safety-relevant load with a power consumption of $P = 1.2\,\mathrm{kW}$ is visualized in Figure 4.12. The safety-relevant systems include all computational units, sensors, and actors required for the driving operation and safety. Further, the light system belongs to this load category.

The entertainment and telecommunication systems are another group of loads sharing a comparable status. Since the passengers directly notice little deviations from the desired state, these systems are excluded from power adjustments through the auction mechanism as far as possible. Hence, the audio system, the interior light, or the screens in the vehicle consume a fixed amount of power based on the user settings and inputs. This power level needs to be considered in the overall auction algorithm but is regarded as a baseline consumption that is not adjustable. Only in rare critical situations in emergency mode, these loads are switched off. Figure 4.13 visualizes the PPF for entertainment and telecommunication systems, which are switched off at the end of the price interval Δp_{crit}.

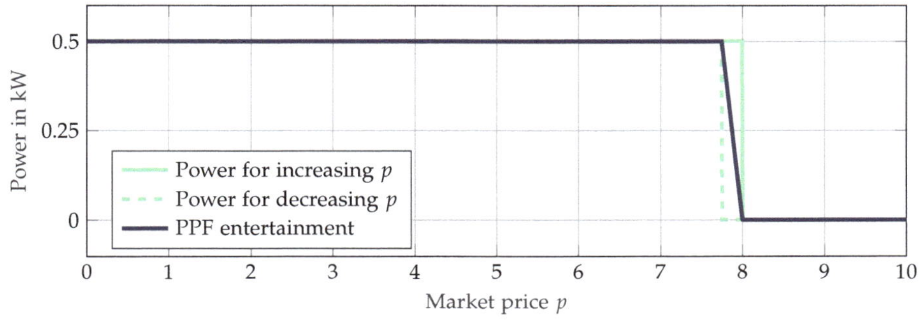

Figure 4.13: PPF design for a telecommunication and entertainment system that consumes a non-negotiable amount of electric power until the critical market price p_{crit} is reached.

4.1.4 Hysteresis Mechanism

Due to the hysteresis for stepwise switching components, such as the seat heating or the blowers, the determined MCP in the power auction may lead to unbalanced power. This is caused by the deviation between the PPFs for the auction mechanism and the actually applied power levels. For the seat heating, these graphs are visualized in Figure 4.10 with the communicated PPF in dark blue and the applied power levels in green. The maximal deviation depends on the difference between the neighboring power levels. Furthermore, this phenomenon increases with every component with a hysteresis in the same price interval.

To provide a suitable foundation for the PES, the proposed power management aims to precisely balance the power supply and consumption. In this regard, the stepwise switching loads and the uncontrollable and highly volatile safety-relevant components pose a great challenge to power management. For a smooth calculation of the MCP in (4.10), the central auctioneer requires a continuous cumulative PPF as depicted in Figure 4.5 [SKSH22]. On the other hand, the physical characteristics and the switching behavior of components and systems in the APN have to be considered. Thus, the hysteresis for seat heating and blower is a trade-off for a smooth MCP determination, the prevention of cyclic switching, and a precisely balanced power in the APN.

To achieve a precise MCP, in [SSH22], a procedure is proposed, which coordinates the various PPFs of the switching components. Within this algorithm, the consideration of user preferences plays an important role. So, components and systems with a higher priority are switched off later than those with a lower user priority.

The primary price range for the hysteresis mechanism is the window for power reduction Δp_{red} where the components decrease their power demand based on the market price p. Some systems or components may have the ability to consume additional power in phases of excessive power availability. Thus, the price range Δp_{rec} is also regarded in the procedure. Regardless of the arrangement, all comfort loads will be switched off if the market price p exceeds p_{crit}.

In order to examine the price intervals for the individual switching actions, all the information about the number and amount of power stages for hysteresis is gathered from the central auctioneer. Furthermore, the user preferences are taken into consideration as additional information for the procedure. With the sum of all switching power stages

$$P_{\text{switch}} = \sum_{i \in \mathbb{S}} P_{\text{max},i} \quad \text{with} \quad \mathbb{S} = \{\text{seat, blower}\} \tag{4.25}$$

and the individual power stages, the relative price ranges are calculated by

$$\Delta p_{\text{blower}} = \frac{P_{\text{blower}}}{P_{\text{switch}}} \quad \text{and} \quad \Delta p_{\text{seat}} = \frac{P_{\text{seat}}}{P_{\text{switch}}}. \tag{4.26}$$

In the exemplary calculation in (4.26), a set of two switching components $\mathbb{S}$ is considered. First, the respective maximal power consumption $P_{\max,i}$ for the components $i \in \mathbb{S}$ is aggregated. While Δp_{blower} and Δp_{seat} describe the intended price intervals relative to the sum of all power stages P_{switch}, the absolute arrangement and order in the price intervals Δp_{red} and Δp_{rec} is examined in accordance to the passengers' preferences. All possible power stages for a component are considered for the distribution of the price segments. If a power stage is missing in a segment due to the user inputs, the PPF is kept constant for that price interval. Hence, the arrangement changes in accordance with the current normal power level set by the user or the operation strategy. Since components directly bordering the price threshold p_{red} are reduced first, components are arranged contrary to the user priorities. Thus, the comfort loads with a low user priority will be switched off first. As mentioned and visualized in Figure 4.13, entertainment systems are put at the end of the price interval Δp_{red} to maintain passenger comfort.

In summary, the following steps are performed within the proposed hysteresis mechanism:

1. Aggregate information about all switching components $i \in \mathbb{S}$ with their power stages and the maximal power consumption $P_{\max,i}$.
2. Divide the price ranges Δp_{rec} and Δp_{red} and distribute the intervals according to the relative price intervals Δp_i and the user priorities.
3. Communicate the absolute price intervals Δp_{hyst} to the components.

To limit the computation and communication effort, the hysteresis algorithm arranges the switching intervals once at the vehicle start. For every recalculation, the information about switching components is aggregated. Thus, the plug-and-play integration or the exclusion of components and systems is ensured. The distributions in the price intervals Δp_{rec} and Δp_{red} differ since not all systems offer the flexibility for additional power consumption during recuperation phases. Hence, only flexible systems are scheduled in Δp_{rec} while every comfort system is included in Δp_{red} for load reduction.

4.2 Predictive Extension Exploiting Flexible Comfort Loads

With the proposed auction-based approach in the previous Section 4.1, the primary goals, namely a capable power balancing mechanism and a flexible power management design facilitating the plug-and-play integration of new components, have been achieved.

The formalized auction-based mechanism comprises the UPA algorithm with an extension for multiple voltage levels, a distributed decision-making and individualization by individual PPFs, and a hysteresis mechanism to smoothly integrate components with switching behavior. Hence, the proposed approach represents a beneficial design for the PMS, enhancing flexibility and offering advantageous characteristics for the development and operation of modern vehicles.

However, the interplay of components and systems in the APN allows for further improvements regarding energy efficiency and the mitigation of aging effects [KKJ+07, RSSH21]. In [RBW+12, LLRB+14, VA15], the idea of utilizing the flexibility of the HVAC system for load shifting is described. For an HEV configuration, the authors in [VA15] propose to exploit the HVAC system's predictive control to counteract the EM's power variations and thereby mitigate the load volatility. As a result, the maximum discharging current of the battery storage is alleviated, which reduces the strain on the battery pack and, thereby, the long-term aging effects.

In [SSH21] and [SKSH22], possibilities for the combination of the formalized auction-based approach and the predictive control are proposed and demonstrated. The advantageous predictive measure and its integration into the auction mechanism are described in the following.

First, the HVAC system is modeled in order to design a suitable MPC algorithm. Since the MPC for the HVAC system makes use of the storage capacities for thermal energy, these storages and the system dynamics are modeled accordingly. Subsequently, the MPC algorithm for the HVAC system is described. Finally, the predictive control approach is integrated into the auction mechanism, combining the complementing features.

4.2.1 Modeling of the Heating, Ventilation, and Air Conditioning

To exploit the HVAC flexibility regarding load shifting, an accurate model including the characteristic time constants and the storage capabilities is required. With the system model and the MPC for predictive operation, the intended functionality with all its constraints can be assured while further goals may be achieved [VA15, BBM17]. To predict and control the future system, the proposed model has to describe the physical behavior appropriately.

Therefore, in [SKSH22], a model focusing on the cabin temperature T_{cab} and the power consumption of the HVAC is presented. The model describes the characteristic temperatures in the system, which give insights into the internal thermal storages. Furthermore, the HVAC power consumption is derived, representing the connection to the auction mechanism for power balancing. More details according to the physical dynamics and design of the HVAC are given in the Appendix A and in [KLFE11, Ste14, Rei14, VA15]. The Appendix A also provides parameters for an exemplary HVAC model implementation.

The HVAC and the thermal system of a vehicle are nonlinear in their dynamics. Thus, a reasoned simplification is necessary to derive a suitable model that depicts a trade-off between the required accuracy and the computational burden. As a result, the following temperatures

$$x(t) = \begin{bmatrix} T_{\text{front}} \\ T_{\text{side}} \\ T_{\text{rear}} \\ T_{\text{body}} \\ T_{\text{roof}} \\ T_{\text{int}} \\ T_{\text{cab}} \end{bmatrix} \quad (4.27)$$

are selected to express the thermal behavior of the HVAC system and, in particular, the vehicle interior. The chosen temperatures include the front windshield T_{front}, the side windows T_{side}, the rear windshield T_{rear}, the vehicle body T_{body}, the vehicle roof T_{roof}, the vehicle interior T_{int}, and the cabin air T_{cab}. As the system states $x(t)$, these temperatures describe the energy storages of different parts and components in the vehicle, which influence the HVAC system operation.

The vector

$$u(t) = \begin{bmatrix} T_{\text{amb}} \\ E_{\text{solar}} \\ T_{\text{pers}} \\ \dot{Q}_{\text{HVAC}} \end{bmatrix} \quad (4.28)$$

provides the model input, which summarizes the external impact and the heat input by the HVAC system. The ambient temperature T_{amb} describes the heat exchange between the vehicle hull and the surrounding air. The solar radiation E_{solar} is an additional heat source for the cabin temperature T_{cab} and the other parts of the interior. With T_{pers}, the temperature of the passengers is given, depicting the heat input through body heat. The controllable actuation of the HVAC system is $\dot{Q}_{\text{HVAC}}$, which is utilized to control the desired cabin temperature $T_{\text{cab,ref}}$.

To describe the dynamic behavior of the thermal system, the derivative

$$\dot{T}_{\text{cab}} = \frac{-\sum \dot{Q}_{\text{conv,in}} + \dot{Q}_{\text{pers}} + \dot{H}_{\text{in}} - \dot{H}_{\text{out}}}{c_{\text{air}} m_{\text{air}}} \quad (4.29)$$

is used. It comprises the internal heat exchange caused by convection $\dot{Q}_{\text{conv,in}}$ (in J/s), the body heat input of the passengers $\dot{Q}_{\text{pers}}$, and the enthalpy from the incoming and outgoing airflow $\dot{H}_{\text{in}}$ and $\dot{H}_{\text{out}}$ (in J/s), respectively. With the thermal capacity of the air

$$C_{\text{air}} = c_{\text{air}} m_{\text{air}} \quad (4.30)$$

in the vehicle, the temperature change of the cabin air $\dot{T}_{\text{cab}}$ in K/s is calculated. The specific convection heat

$$\dot{Q}_{\text{conv,i}} = \alpha_i A_i (T_{\text{cab}} - T_i) \quad (4.31)$$

is expressed by the temperature difference, the surface A_i, and the individual coefficient of heat exchange α_i. By the product of mass flow $\dot{m}_i$ and the specific enthalpy h_i, the enthalpies $\dot{H}_{\text{in}}$ and $\dot{H}_{\text{out}}$ are derived. It is assumed that the incoming and outgoing air masses $\dot{m}_{\text{in}}$ and $\dot{m}_{\text{out}}$ are equal and, thus, the enthalpy difference in (4.29) can be simplified by

$$\Delta\dot{H} = \dot{H}_{\text{in}} - \dot{H}_{\text{out}} = \dot{m}_{\text{in}}(h_{\text{in}} - h_{\text{out}}).$$

The enthalpy difference $\Delta\dot{H}$ represents the heat input of the HVAC system $\dot{Q}_{\text{HVAC}}$, which is included in the input vector (4.28). The temperature derivatives of the other components i are expressed by

$$\dot{T}_i = \frac{\dot{Q}_{\text{conv, in}} + \dot{Q}_{\text{conv, amb}} + \dot{Q}_{\text{solar}} + \dot{Q}_{\text{rad}}}{C_i}$$

and include the additional impact factors, solar heat input $\dot{Q}_{\text{solar}}$, the heat exchange with the ambient $\dot{Q}_{\text{conv, amb}}$, and the heat exchange by radiation $\dot{Q}_{\text{rad}}$. The thermal capacity for a component is similar to the thermal capacity of the air and is calculated by

$$C_{\text{i}} = c_{\text{i}} m_{\text{i}}. \tag{4.32}$$

The exchange with the surrounding air $\dot{Q}_{\text{conv, amb}}$ through convection is described in (4.31). The heat impact from the solar radiation $\dot{Q}_{\text{solar}}$ and the heat transfer which depends on the radiation mechanism $\dot{Q}_{\text{rad}}$ are given by

$$\dot{Q}_{\text{solar}} = \epsilon_{\text{rad}} A_i E_{\text{solar}}$$

and

$$\dot{Q}_{\text{rad}} = \epsilon_{\text{rad}} \sigma_{\text{rad}} A_i (T_i^4 - T_{\text{amb}}^4), \tag{4.33}$$

respectively. The coefficient ϵ_{rad} specifies the characteristic absorption of the solar radiation for a particular surface. For the general radiation term in (4.33), the factor σ_{rad} represents the Stefan-Boltzmann constant.

In summary, the introduced equations are a simplified mathematical representation of the physical HVAC system behavior. The model explains the relationship between the heat input from various sources and the mechanism of the HVAC system, heating or cooling the cabin temperature T_{cab}. Consequently, the model provides a connection between the cabin temperature T_{cab} and the power consumption of the HVAC system in terms of the controllable heat exchange $\dot{Q}_{\text{HVAC}}$. In the next section, an MPC algorithm is designed based on the temperature model and the two contradicting goals, temperature control and flexible load shifting.

4.2.2 Model Predictive Control for the Heating, Ventilation, and Air Conditioning

To control the desired cabin temperature T_{cab} while considering various constraints, model predictive control (MPC) is a suitable control methodology, which is widely applied in industry and science [BBM17]. Since the maturity of MPC theory is high, there is sufficient information about the fundamentals and the specific design for different systems. Hence, a thorough introduction is omitted in this work. For further insights, it is referred to the literature [KC16, BBM17].

In general, MPC algorithms comprise the following characteristics [KC16, BBM17, RSSH21]:

1) A dynamic optimization problem is defined and solved in every new time step, yielding the next control input.
2) The required prediction of the targeted system makes use of an internal model to describe the future system behavior.
3) Constraints regarding the inputs, the states, and the outputs can be taken into account during the optimization procedure.
4) Regarding a moving horizon, the MPC applies the first control input of the optimized input trajectory to the system while discarding the following input values. Afterward, the system output is measured or estimated in order to adapt the control strategy iteratively.

An advantageous and valuable characteristic of MPCs is the seamless consideration of constraints. Additionally, the adjustment of the moving horizon allows for consideration of information reliability and computational power. Hence, the MPC is adaptable according to the prediction horizon, in which information is reliable. On the other hand, the moving horizon length directly correlates with the computational effort regarding the dynamic optimization problem.

The proposed MPC algorithm for the HVAC system controls the cabin temperature T_{cab} with respect to the user reference $T_{\mathrm{cab,ref}}$. The objective function

$$J = \sum_{k=1}^{N_p} \left(T_{\mathrm{cab}}(k) - T_{\mathrm{cab,ref}}(k)\right)^2$$

expresses the goal regarding the prediction horizon N_p. During the optimization procedure, the MPC makes use of the previously derived HVAC and vehicle interior model to predict the system behavior and the cabin temperature T_{cab}. On the other hand, the MPC is capable of considering additional constraints, such as the power limits that result from the request to shift load away from peak power phases.

4.2.3 Integration into the Auction Mechanism

To smoothly integrate the predictive extension into the auction-based approach from Section 4.1, the accumulated demand and supply of the power network for the proceeding time horizon N_p are determined. The resulting cumulative supply and demand $\mathrm{PPF}_{\mathrm{cum}}$ for an instant of time k is highlighted in Figure 4.14. It is assumed that the aggregated supply and demand for the next time steps within the prediction horizon N_p are given by $\mathrm{PPF}_{\mathrm{cum}}(p_k)$ with $k \in \{1, \dots, N_p\}$. The various electric and electronic components in the APN have to provide the necessary information to calculate $\mathrm{PPF}_{\mathrm{cum}}$ for the next time interval $k \in \{1, \dots, N_p\}$. This may be done by modeling and prediction or an extrapolation based on the current consumption or generation.

As long as the price p is in the normal price range $\Delta p_{\mathrm{norm}} \in (p_{\mathrm{norm}} = 2, p_{\mathrm{red}} = 6]$, there is no need for load shifting or similar measures. Hence, all components and systems optimize the individual operation regarding efficiency and effectiveness. If the predicted prices p_k in the prediction horizon N_p leave the normal price range Δp_{norm}, countermeasures are taken by the predictive extension.

For a price beyond p_{red}, the availability of power in the APN is unsatisfying. Thus, the predictive extension algorithm requests flexible loads to shift load demands, mitigating the respective power peak. If there is a surpassing power availability indicated by a market price $p \leq p_{\mathrm{norm}} = 2$, an additional load demand is required to fill the gap.

Due to the simple consideration of constraints in the MPC design, the depicted countermeasures regarding load shifting are applied by time-varying power limits for the flexible loads. Therefore, a lower and upper limit are derived by

$$P_{\mathrm{up}} = \mathrm{PPF}_{\mathrm{cum}}(p_{\mathrm{red}}) \quad \text{and} \quad P_{\mathrm{low}} = \mathrm{PPF}_{\mathrm{cum}}(p_{\mathrm{norm}}), \tag{4.34}$$

with p_{red} and p_{norm} as the market price boundaries for the upper and lower power level in Figure 4.14. Additionally, the power reserves between the power balance

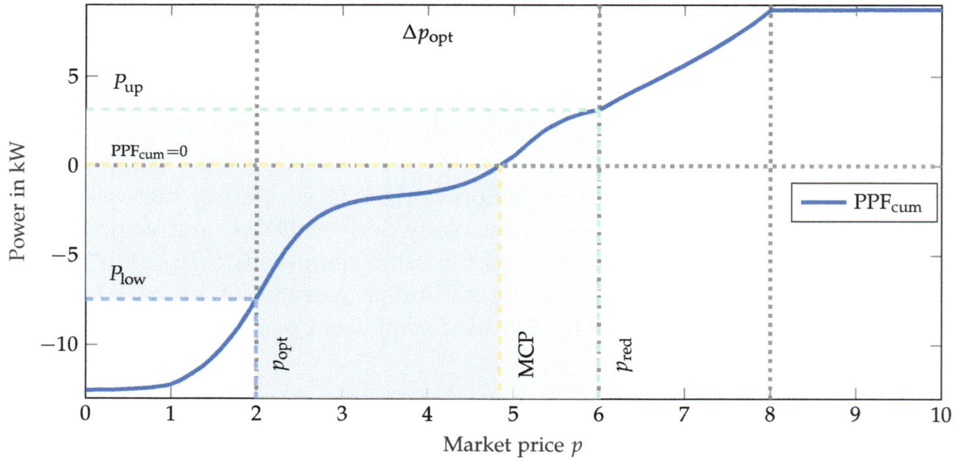

Figure 4.14: Cumulative PPF with the MCP for $\mathrm{PPF}_{\mathrm{cum}} = 0$, the upper power limit P_{up}, and the lower power limit P_{low} within the market price interval Δp_{norm}.

marked by $\mathrm{PPF}_{\mathrm{cum}} = 0$ and the upper limit P_{up} and the lower limit P_{low} are highlighted. The flexible extension aims to keep the market price p inside the normal and desired operation range Δp_{norm}, which comprises the light green and blue areas. For the proceeding prediction horizon, the power limits are given by

$$P_{\mathrm{up}}(k) \text{ and } P_{\mathrm{low}}(k) \quad \text{with} \quad k \in \{1, \ldots, N_p\}. \tag{4.35}$$

The predictive extension passes the calculated power limit predictions in (4.35) to the predictive control algorithms of flexible loads. For an exemplary APN, Figure 4.15 depicts the overall power limits for a time interval of fifteen seconds. Generally, the upper power limit P_{up} in green is above zero, and the lower power limit P_{low} in blue is below zero. As highlighted in Figure 4.14 and in Figure 4.15, in these situations, the MCP in yellow is in the normal interval Δp_{norm}. If the upper limit falls below zero, the normal price band Δp_{norm} is left due to an unsatisfying power supply as shown in the detail view in Figure 4.16a for the time interval from zero to five seconds. On the other hand, if the lower limit P_{low} rises above zero, there is a surpassing power supply, which may lead to strain on the battery storage due to excessive charging currents. In Figure 4.16b, the lower limit P_{low} is above $\mathrm{PPF}_{\mathrm{cum}} = 0$ for about two seconds. However, in both situations, there are phases with power reserves marked by the green and blue areas. The predictive extension utilizes these power reserves to shift the power demand of flexible loads accordingly.

Since the limits in (4.35) refer to the complete APN, there is an adaption needed to apply these limits as constraints for the HVAC system's MPC. Hence, the limits are adjusted by

$$\overline{u}_{\mathrm{power}} = \min\{P_{\mathrm{HVAC,max}}, P_{\mathrm{up}}(k) + \hat{P}_{\mathrm{HVAC}}(k)\} \tag{4.36}$$

and

$$\underline{u}_{\mathrm{power}} = \max\{P_{\mathrm{HVAC,min}}, P_{\mathrm{low}}(k) + \hat{P}_{\mathrm{HVAC}}(k)\} \tag{4.37}$$

with $k \in \{1, \ldots, N_p\}$. The estimated HVAC consumption $\hat{P}_{\mathrm{HVAC}}$ is extracted from the aggregated power P_{up} and P_{low} to get the power limit from the HVAC perspective. In this regard, the adjusted power limits are equivalent to an adapted cumulative PPF with

$$P_{\mathrm{up}}(k) + \hat{P}_{\mathrm{HVAC}}(k) = \sum_{i \in \mathrm{APN} \setminus \{\mathrm{HVAC}\}} \mathrm{PPF}_i(p_{\mathrm{red}}) \tag{4.38}$$

and

$$P_{\mathrm{low}}(k) + \hat{P}_{\mathrm{HVAC}}(k) = \sum_{i \in \mathrm{APN} \setminus \{\mathrm{HVAC}\}} \mathrm{PPF}_i(p_{\mathrm{norm}}). \tag{4.39}$$

Additionally, in (4.36) and in (4.37), the absolute power limits of the HVAC system $P_{\mathrm{HVAC,min}}$ and $P_{\mathrm{HVAC,max}}$ are incorporated since these are preset by design and restrict the inherent flexibility.

Reliable information about the future demand and supply is needed for a suitable integration of the extension. Thus, the various components and systems in the APN

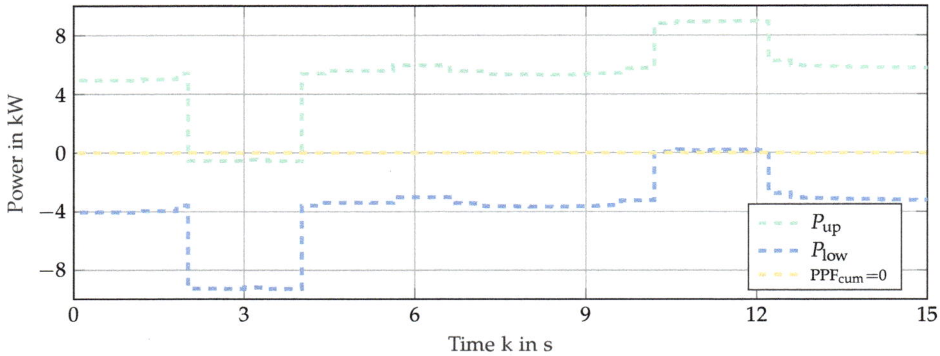

Figure 4.15: Power limits determined in (4.35) with the market price p_{norm} for the lower limit P_{low} and p_{red} for the upper limit P_{up} (see Figure 4.5). The predictive extension intervenes if the upper limit P_{up} is below or the lower limit P_{low} is above $\mathrm{PPF}_{cum} = 0$, respectively.

that take part in the auction have to predict their individual PPFs for the prediction horizon N_p.

Currently, the possibilities and implementations for estimating future consumption in the vehicle are rare since there is little focus on this area [HWK+17, RSSH21]. For EVs, the estimation of the remaining range is an important field of research because consumers feel unsafe about the range estimation in EVs [PTS+19]. However, these estimations are derived from previous rides and put emphasis on the mean value rather than the exact power consumption estimate in the next seconds or for the remaining ride.

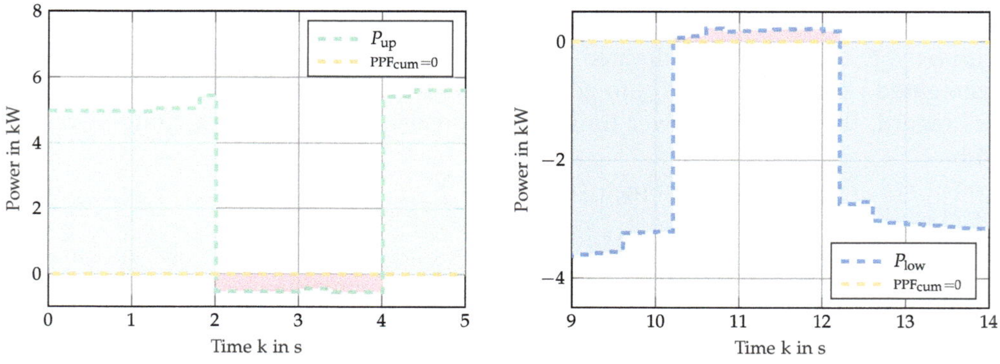

(a) Detailed view on P_{up}.

(b) Detailed view on P_{low}.

Figure 4.16: Detailed views of the power limits activating the predictive extension if the normal price interval Δp_{norm} is left. The red areas indicate the lack in Figure 4.16a or the surplus in Figure 4.16b of power in the APN.

Nevertheless, information about the driving situation and the surroundings is available, which provides the basis for the required power estimation. One crucial part is the driving operation, representing the cause for the highest power peaks through boosting or recuperation by the EM. Here, information from the navigation system comprising data about road characteristics, speed limits, road slope, or even the predicted speed profile are present in modern vehicles [LLRB+14, HWK+17]. In [KGT19], the prediction of driving loads is implemented by the utilization of neuronal networks.

By incorporating weather conditions from external information and temperature information from internal sensors, the HVAC system, as another major power consumer, is capable of predicting the power consumption for cooling or heating operations. The dissemination of autonomous driving functionalities in the near future opens up further possibilities since the planned driving trajectory is available beforehand and may even be adaptable to optimize overall performance [VA15, HWK+17].

For an exemplary scenario, the constraints $\overline{u}_{\text{power}}$ in green and $\underline{u}_{\text{power}}$ in blue for the HVAC power are depicted in Figure 4.17 for a five-second time interval. The inherent HVAC system power limits are set to $P_{\text{HVAC,min}} = 0\text{W}$ and $P_{\text{HVAC,max}} = 3500\text{W}$. The constraints $\overline{u}_{\text{power}}$ in green and $\underline{u}_{\text{power}}$ in blue result from the calculations in (4.36) and (4.37). Hence, compliance with these inherent HVAC power limits ensures that the market price p is kept within the normal price range Δp_{norm}. In this regard, the dark blue dashed graph for the HVAC with the predictive extension in Figure 4.17 demonstrates the adaption to the prevailing power constraints.

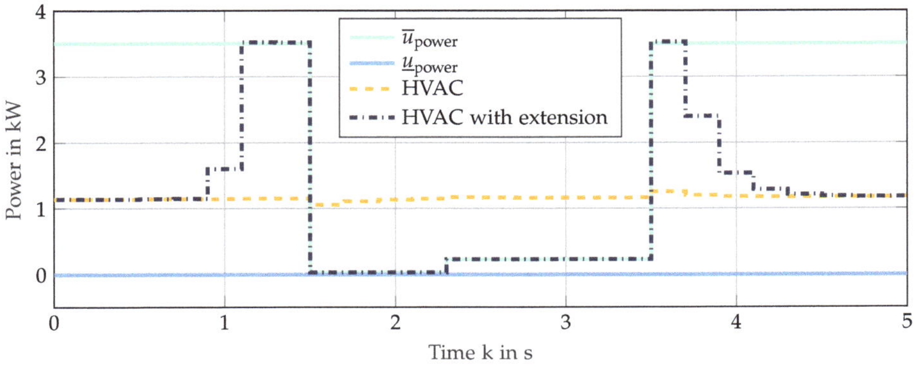

Figure 4.17: Input constraints of the HVAC system calculated in (4.36) for the upper constraint $\overline{u}_{\text{power}}$ and in (4.37) for the lower constraint $\underline{u}_{\text{power}}$. The model predictive control ensures compliance with the respective power constraints.

4.3 Communication based on a Service-Oriented Architecture

The proposed auction-based power management offers flexibility and the possibility for plug-and-play integration of new components, seamlessly integrating into the group of auction participants. With the complementing predictive extension, the algorithm is further improved according to a predictive power balance and the exploitation of flexible loads in the APN.

An essential aspect of the implementation and operation is the communication scheme. Since the service-oriented architecture (SOA) is an emerging technology in the automotive industry, current vehicles and future developments in communication architecture are based on the SOA [HSS+22, SSG+22]. In contrast to the classic signal-based communication, the SOA facilitates the configuration of communication paths at vehicle start or even at runtime [RGKS20]. By the communication pattern publish and subscribe, functions, called services in the SOA, get access to various information in the vehicle. Therefore, the aim is to develop a communication design with the potential to seamlessly withdraw or integrate new hardware components into an existing APN [BW09].

While providing the desired flexibility in communication for comfort and entertainment systems, there are challenges applying the SOA in safety-relevant functionalities, for instance, autonomous driving functions [SSG+22, HSS+22]. For detailed information about the working principle, the advantages, and the current research and applications, it is referred to [KOB+17, RGKS20, HSS+22].

Communication Overview

Concerning the proposed auction-based power management and the predictive extension, various information have to be communicated between the participants at certain times or in a cyclic manner [SSH22]. Table 4.2 and Table 4.3 list the different communication paths with the respective sender, receiver, content, and frequency of the communication.

In terms of frequency, there is information sent once at the vehicle start, information sent with every change, and information sent frequently with a fixed cycle time ΔT_k. Here, ΔT_k represents the auction mechanism's cycle time, which is performed periodically to coordinate the power consumption and supply in the APN. In this work, all information is received via the publish and subscribe pattern.

For the initialization at the vehicle start, the individual PPFs have to be known. For vehicles with multiple voltage levels $m \in \mathbb{V}$, the voltage level membership of all components and systems in the APN has to be communicated. With the voltage level membership information, $\mathrm{PPF}_{\mathrm{cum},m}$ with $m \in \mathbb{V}$ for the voltage levels are calculated to monitor the power transfer limitations $P^{lm}_{\mathrm{PE,lim}}$ between the connected power levels l and m.

Another aspect of the communication strategy is the implementation of the proposed hysteresis mechanism. To reduce the communication overhead and prevent failures in the auction algorithm during operation, the mechanism for the hysteresis distribution is performed with every vehicle start. For the relative power distribution in the price intervals Δp_{rec} and Δp_{red}, the power stages of all components with switching characteristics and the information about flexible power consumption are required. Furthermore, the user preferences are included to examine the absolute arrangement in the price intervals Δp_{rec} and Δp_{red}. Subsequently, the individual price intervals Δp_{hyst} are sent to the respective components.

Regarding the communication at runtime, information is sent with a change or periodically with the cycle time ΔT_k. To determine the explicit PPFs, the auctioneer needs the information about the component states. Furthermore, the auctioneer receives the current power transfer limitations $P^{lm}_{\mathrm{PE,lim}}$ to consider these for the next auction. Finally, subsequent to the periodic UPA, the MCP or the individual market prices p^m_k for the voltage levels m are forwarded to the components, which adapt their power consumption accordingly. The auctioneer sends the power transfer commands P^{lm}_{ref} to the respective DC/DC-converters, implementing the power transfer between voltage levels.

The predictive extension uses information from further internal and external sources. Nevertheless, it is assumed that most information is processed within the individual components and, thus, not considered in Table 4.3. As a basis for the predictive extension, the auctioneer requires future load profiles, which are depicted by the individual PPFs design of the components in Table 4.2. To predict the future supply or consumption of power, data about internal system states, sensor information, or information from the navigation system are assessed and processed for the prediction

Table 4.2: Communication between the different participants in the auction mechanism according to the proposed communication concept.

Sender	Receiver	Content	Frequency
Components	Auctioneer	PPF design	At vehicle start
Components	Auctioneer	Voltage level membership	At vehicle start
Components	Auctioneer	User preferences	At vehicle start
Switching components	Auctioneer	Load steps for hysteresis	At vehicle start
Components	Auctioneer	Information and states	With change
DC/DC-converter	Auctioneer	Power transfer limitations	With change
Auctioneer	Switching components	Hysteresis design	At vehicle start
Auctioneer	Components	Market price p	ΔT_k
Auctioneer	DC/DC-converter	Power transfer commands P^{lm}_{ref}	ΔT_k

Table 4.3: Additional communication between the different participants regarding the predictive extension of the auction-based power management.

Sender	Receiver	Content	Frequency
Components	Auctioneer	Predicted information and states	With change
DC/DC-converter	Auctioneer	Predicted power transfer limitations	With change
Auctioneer	HVAC	Power constraints for MPC	ΔT_k

horizon N_p.

With the calculation of the power limits, the auctioneer monitors whether the market price remains in the normal price range Δp_{norm} or not. If the price drops out of this range, power constraints are communicated to the respective flexible load, for instance, the HVAC system. The communication with the involved flexible load is continued until the desired price range Δp_{norm} is maintained without load-shifting measures. In comparison to the auction-based approach, the communication comprises predicted information about the component states and the time-varying power limits for the flexible loads.

Centralized and Distributed Auction Design

The elaborated communication overview provides information about data contents and the participants in the proposed auction-based mechanism. Within the design principles of the SOA, there are various possibilities to implement the auction mechanism. In the following, two approaches are presented, which bring specific advantages or disadvantages regarding communication effort and fault tolerance.

The first communication design focuses on *centralized* processes and calculations. Accordingly, the main parts of the auction-based algorithm are applied in a central computational unit that represents the central auctioneer. Table 4.2 and Table 4.3 refer to the centralized implementation. The overview in Figure 4.18 highlights the communication paths between the auction participants for this particular implementation. In order to determine the MCP for the next time step k, the auctioneer depends on the information about the current PPFs. Thus, the auctioneer collects the design patterns of the PPFs and the information about component states as summarized in Table 4.2. Consequently, the central unit calculates the MCP internally and communicates it to the incorporated components, yielding a balanced power in the APN.

Since the publish and subscribe pattern in the SOA offers the possibility to spread general information easily, the central auctioneer service gathers available information by subscription. Hence, most state information, for example, the SOC of battery storage, is already accessible in the vehicle. Further, the notification of the next market

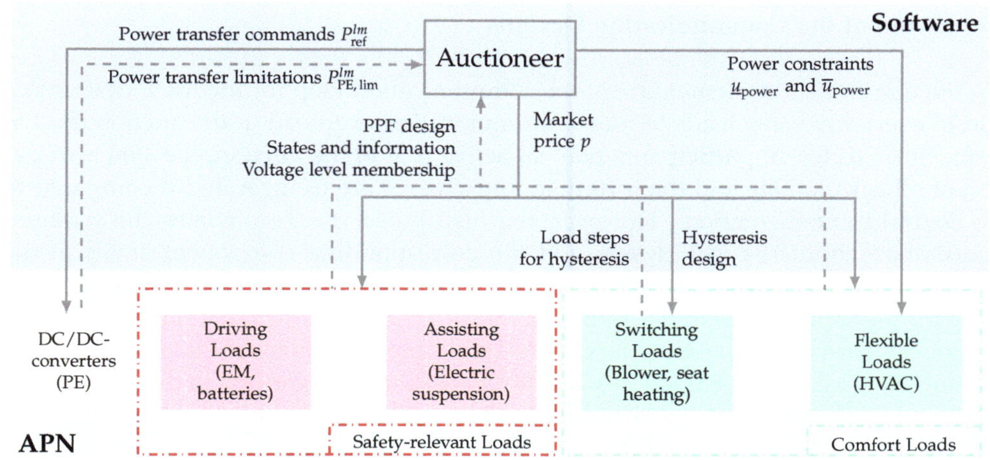

Figure 4.18: Communication architecture regarding the proposed auction-based power management and the predictive extension concerning flexible loads. For further details regarding the different loads, see Chapter 3 and Figure 3.8.

price p_k is a service that all corresponding participants in the auction subscribe to and that is updated periodically.

For the calculation of the current MCP and the future MCPs concerning the prediction horizon N_p, all computation is performed in the central auctioneer. As a result, it is easier to ensure the MCP determination on time since all computation is clustered in one ECU. Apart from the technical benefits or drawbacks of the centralized auction design, the centralization in vehicle architecture regarding information and computation power is an ongoing development [BLOT19, ORBE20].

In contrast to the centralized design, the *distributed* implementation conducts the MCP determination in a comprehensive loop, which comprises the auctioneer as well as all auction participants. To examine the MCP through market price adjustments, the current market price p_k is sent to all participants. As an answer, the participants send their power offers or bids to the auctioneer as depicted in Figure 4.1. The auctioneer aggregates the offers and bids to adjust the price according to the rules in Definition 4.5. This procedure is performed until the power balance in the APN is reached. As a result, the adaptation loops in Figure 4.6 include all auction participants.

Subsequently, the final market price (MCP) is communicated to the APN components so that all auction participants adapt their behavior. For the predictive extension, the described loop has to comprise a vector of market prices p_k, and all predicted offers and bids for $k \in [1, N_p]$. Hence, all predicted MCPs and the power limits for the flexible load are calculated simultaneously in the comprehensive loop.

Discussion of the Communication Designs

For the distributed implementation, the communication loop for the MCP determination in every time step leads to more communication overhead in the auction mechanism. Since different participants play an active role in the auction, the synchronization of all calculations and the information exchange are exaggerated if compared to the central implementation. However, the distributed design obviates the required information about the PPF design and the corresponding component states in the PPFs. Consequently, the distributed adjustment of the individual PPF leads to flexibility for the components in the APN and less information accumulation in the central auctioneer service. In contrast, the centralized implementation facilitates a faster and deterministic MCP determination since all calculation is performed in the central auctioneer unit. Thus, the MCP determination does not rely on a robust and deterministic communication protocol.

In summary, there are different possibilities for communication design. The decision about the communication design plays an essential role in the implementing the proposed auction-based power management. With the iterative reinitialization at vehicle start, the proposed mechanism allows for the seamless integration or the removal of components and the implementation of new user preferences and priorities [Ise16, p. 98]. Thus, the SOA properties complement the functional advantage of the proposed auction-based power management, leading to a seamless plug-and-play integration for new hardware components.

Another aspect that needs to be emphasized is the increased fault tolerance facilitated by the SOA. Since there is a flexible service allocation among the computational units in the vehicle at runtime, the central auctioneer as a service may be instantiated on different ECUs to increase the fail-safety [GYZ+10, SGLS22, HSS+22]. If the active auctioneer instance is out of order, the fallback auctioneer service takes over the critical role. The fallback auctioneer service is the so-called shadow of the primary auctioneer service. It aggregates the same information and seamlessly overtakes the action if the primary auctioneer is unavailable. Hence, the SOA design ensures the operation of the PMS for single points of failure with regard to the central auctioneer.

4.4 Conclusion

This chapter presents the predictive auction-based automotive power management and the fruitful combination with the SOA. In Section 4.1, the auction mechanism is formalized with the rules in Definition 4.1 for a reliable auction framework. The approach comprises the generalization toward multiple voltage levels, the individualization with PPFs, and the hysteresis mechanism for switching loads. In summary, the proposed power management facilitates a high quality of power balancing in the

APN, consideration of individual component behavior, and the effortless integration of new hardware components as participants in the auction.

The predictive extension in Section 4.2 supports the consideration of additional optimization goals. In this regard, the purposeful load shifting concerning flexible comfort loads enables an optimized power distribution over time and, for example, mitigates strain on the battery storage.

The elaborated communication designs in Section 4.3 depict different opportunities for the auction implementation in terms of the necessary communication paths. Applying the SOA yields advantages, such as a fail-safe design with multiple service instances for the central auctioneer. In a nutshell, the proposed automotive power management is a promising innovation for future APNs, offering great flexibility in vehicle development and operation.

5 Auction-based Automotive Power Management: Simulations and Results

This chapter presents the simulation results regarding the proposed auction-based power management and the predictive extension. To focus on the fundamental working principles, an ideal communication without false information or delayed transmission is assumed. Therefore, the simulations do not consider the implementation and simulation of the communication processes via the SOA, as well as the possible uncertainties and disturbances in the load prediction.

Due to the few research works concerning automotive power management, there are no reference scenarios in terms of APN setup or power profiles. In [Gra04a, Büc08], the authors emphasize the auction-based concept and its adaptation to the APN and energy management. Consequently, this is the first time that auction-based automotive power management has been formulated and evaluated. In this regard, the following simulation results are the initial step toward the validation process, focusing on the mechanism and working principle.

First, the general working principle of the auction-based approach from Section 4.1 is elaborated for a given HEV power network configuration. Section 5.1 comprises the market price adaptation procedure for APNs with multiple voltage levels. In Section 5.2, the simulation results for a driving scenario with a critical power shortage are presented. Subsequently, the influence of stepwise switching loads on the achieved power balance is shown in Section 5.3. In Section 5.4, the effects of variations regarding the individual PPFs are presented for the battery PPF and a changing battery SOC. Finally, the successful plug-and-play integration of new components into the APN and the inclusion into the auction mechanism is demonstrated in Section 5.5.

The predictive extension for the exploitation of flexible comfort loads from Section 4.2 is validated in Section 5.6. The simulation results illustrate the optimization of the cabin temperature via the MPC with prevailing side goals taking into account the need for load shifting in the APN. Furthermore, the load-shifting capabilities of the HVAC system and its limitations are highlighted. In Section 5.7, the simulative results and the approach's advantages and disadvantages are summarized.

Scenario Descriptions

All simulations build upon the given APN in Figure 5.1. The depicted APN comprises two voltage levels $\mathbb{V} = \{12\text{V}, 48\text{V}\}$, which are connected with a DC/DC-converter marked as PE. In the assumed configuration, the APN is part of a HEV with multiple voltage levels for increased reliability and higher efficiency [KET+11, ZZL+21]. Thus, the EM provides the HEV operation modes boosting in (4.21), adaptive power generation in (4.20), and recuperation in (4.22), which are all depicted by the respective PPF in Figure 4.7. The standard operating mode of the EM is the adaptive power generation, in which the EM supplies the power network with the necessary power. Since the EM in the HEV configuration occasionally supports the driving operation, the EM's role in the auction mechanism changes from mostly power selling to power bidding occasionally. Additionally, in recuperation mode, the EM generates high amounts of power to convert kinetic energy into electric energy, thereby increasing the efficiency of the driving operation. As a result, the HEV configuration puts additional stress on the power infrastructure and the desired power balance in the APN.

Figure 5.1 depicts a HEV power network with two voltage levels and one battery storage in each level. The HVAC system in the 48V level includes different components and loads that control the cabin temperature T_{cab} as well as connected states, such as the humidity in the cabin. The behavior of the HVAC system within the auction is determined by the PPF in (4.24). As elaborated in Section 2.1, the electric storage systems are the essential components for passive power balancing in the APN. The batteries supply the vehicle and its electric systems with energy and simultaneously balance power variations in a passive manner. Based on their physical characteristics, the battery storages adapt their behavior in response to the current voltage level. If the voltage level drops due to increased power consumption, the battery provides power while the internal battery voltage is adjusted accordingly. The battery storage is charged if the voltage rises above the open circuit voltage V_{OC}. Hence, besides the active components in the energy infrastructure, such as the DC/DC-converter (PE) or the EM, the batteries naturally balance the power variations. The individual battery

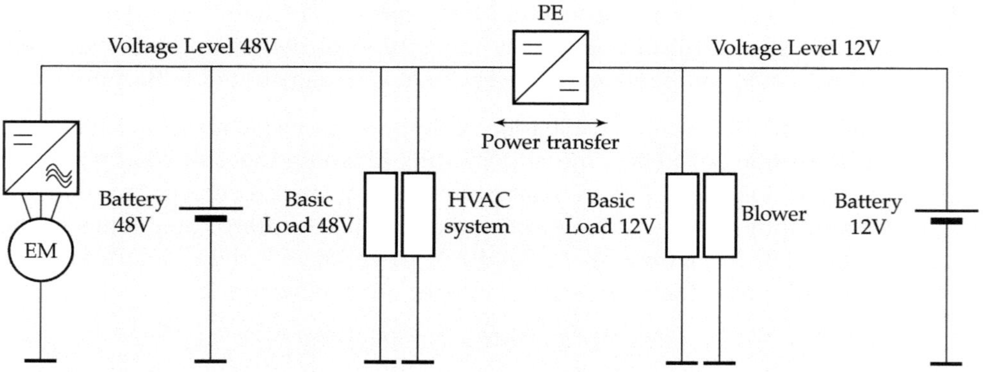

Figure 5.1: HEV configuration with two voltage levels, 12V and 48V, and respective components for the simulations. Compared to Figure 1.1, the loads in the two voltage levels are slightly changed.

chemistry and resulting characteristics are not part of the simulation but may be considered in the individual PPF design described in Section 4.1. A further aspect is the aging of battery cells due to exaggerated use and thermal overloading, described by the SOH. To keep the aging of battery cells at a reasonable level, measures have to be applied to reduce the strain on the battery storage [MOG+09, VA15, RSSH21].

The blowers are situated in the 12V level. The blowers are defined as stepwise switching loads, putting additional challenges on the auction mechanism and the infrastructure. However, the caused power steps are mainly balanced by the battery storages or inherent capacitive elements in the short term. The power balance may be reestablished by the PE or the EM as the active power infrastructure components in the following step. The different mechanisms regarding short-term and long-term power balance are described in Chapter 3.

Additionally, basic loads are placed in both voltage levels. These basic loads incorporate further components not actively participating in the auction mechanism since they are not controllable by the PMS. In this group of loads, there are mainly safety-relevant systems, such as electric steering or suspension. Moreover, entertainment or telecommunication systems are included in this group. Since passenger comfort and safety strongly rely on these systems, they are only deactivated or switched off in rare situations [KET+11].

Driving Scenarios

The two simulated driving scenarios illustrate exemplary situations in the summer season with cooling and air conditioning activities by the HVAC system and the blower. Each scenario comprises a time interval of five seconds. The fundamental simulation design is similar to [SRSH21, SKSH22, SSH22, SGH24] and extended to show more aspects and properties of the proposed auction mechanism.

Each scenario depicts an exemplary situation in which power management is requested to balance power demand and supply. The power imbalance during the first simulation is mainly caused by the EM boosting action, which leads to an insufficient power supply. In the second situation, the EM is in recuperation mode and converts high amounts of kinetic power into electric power, resulting in an exaggerated power availability in the APN.

The power consumption profiles of the basic load 12V and the basic load 48V are visualized in Figure 5.2 and Figure 5.3 for the first and second scenario, respectively. The basic loads are regarded as safety-relevant components and, thus, do not participate in the auction mechanism, resulting in a fixed and uncontrollable power consumption. Besides the depicted load profiles in Figure 5.2 and in Figure 5.3, which refer to less volatile consumers, other safety-relevant loads, such as the electric suspension or the steering motor, cause high peak power demands. These short-term peak power demands are out of the scope of automotive power management and, thus, are not considered in the following simulations. Nevertheless, these peak power demands have to be considered by a possible peak power reserve. The velocity profiles for both simulation scenarios are shown in Figure 5.4.

Due to the high ambient temperature $T_{\text{amb}} = 32\,°\text{C}$, the HVAC system is activated to cool the vehicle. Additionally, the blower is switched on with a power consumption of $P_{\text{Blower},4} = 800\,\text{W}$ in the first scenario and $P_{\text{Blower},1} = 200\,\text{W}$ in the second scenario. The respective basic load in the 12V and 48V levels adds further power demands to the system. On the other side, the EM, in generator mode, and the batteries provide the required power in the APN.

In the first scenario, which is visualized in Figure 5.5, the EM changes to boost mode in order to support the driving operation and accelerate the HEV. As a result, the power balance in the APN alters significantly. At the simulation start, the EM is the primary power supplier. Due to the subsequent change to boost mode after 1.5 s, the EM is the main consumer. The boosting phase ends after 3.5 s and is followed by a

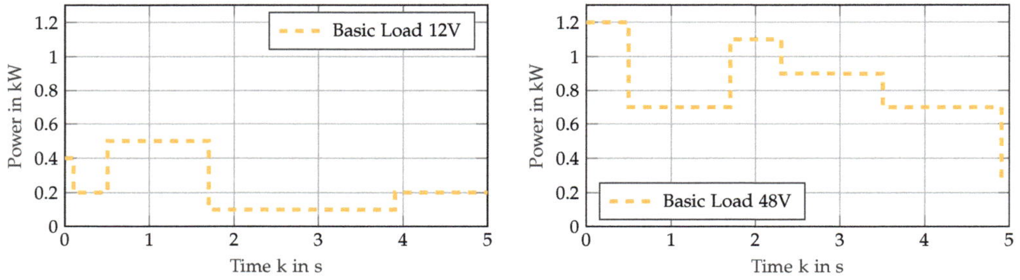

(a) Basic load profile 12V.
(b) Basic load profile 48V.

Figure 5.2: Basic load profiles for both voltage levels and the first scenario with the EM in boost mode. The basic load profiles are non-negotiable within the auction and pose a constant load on the power supply.

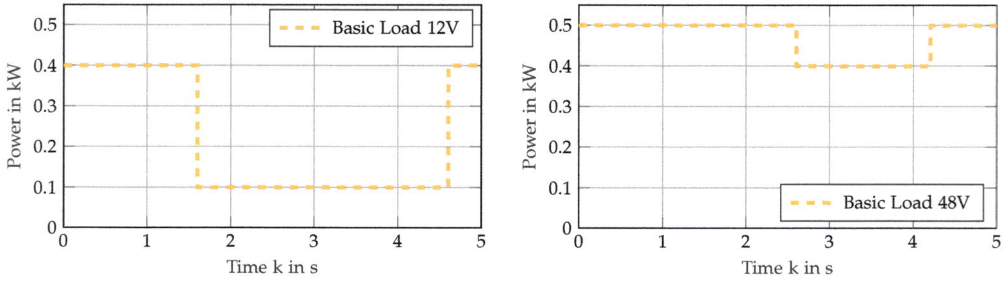

(a) Basic load profile 12V.
(b) Basic load profile 48V.

Figure 5.3: Basic load profiles for both voltage levels and the second scenario with the EM in recuperation mode. The basic load profiles are non-negotiable within the auction and pose a constant load on the power supply.

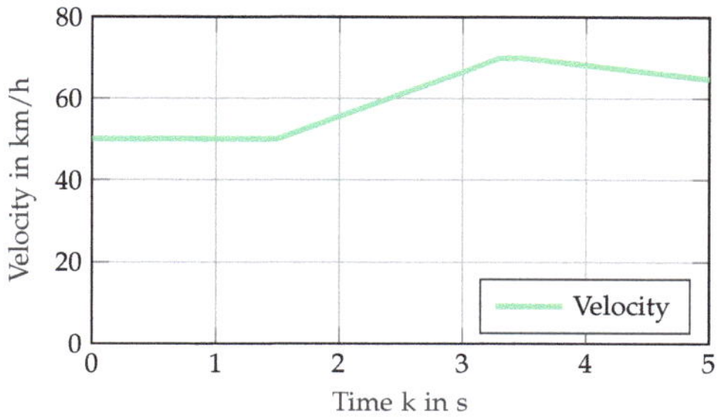

(a) Velocity profile first scenario.

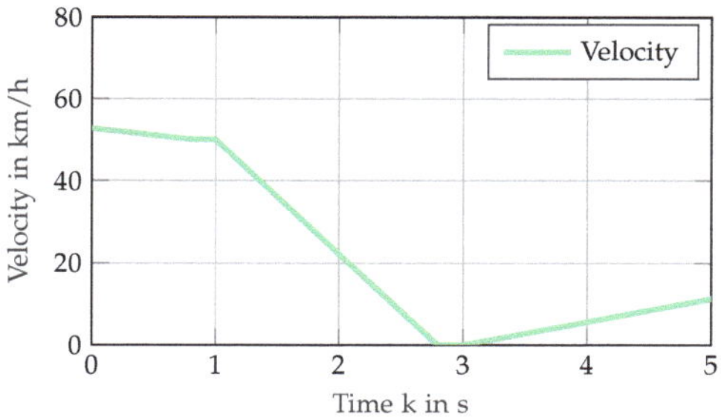

(b) Velocity profile second scenario.

Figure 5.4: Velocity profiles of the first and second scenario determining the EM mode during the scenarios.

period in which the EM is again in generator mode. In this phase, the HEV holds its velocity and slowly decelerates, approaching an intersection with traffic lights. The according velocity is displayed in Figure 5.4a. Since the vehicle velocity is a crucial factor regarding the passive cooling of the vehicle, the HVAC system behavior strongly depends on the current velocity.

In the second scenario, visualized in Figure 5.6, the EM is in recuperation mode, transforming braking energy into electric energy. The HEV stops in front of the traffic lights, and the velocity is reduced to zero, which is depicted in Figure 5.4b. Subsequently, the EM switches to generator mode for the remaining simulation time. At the green signal, an acceleration phase is conducted from $k = 3\,\mathrm{s}$ to $k = 5\,\mathrm{s}$. The mechanical power for the acceleration is solely provided by the ICE.

The corresponding parameters and state information for both scenarios are listed in Table 5.1. The HVAC model in Section 4.2 is utilized to calculate the desired power of the HVAC. Through the HVAC operation, the cabin temperature T_{cab} is kept in the desired range around $T_{\mathrm{cab,\,ref}} = 21\,°\mathrm{C}$. In this work, it is assumed that a range of $1\,°\mathrm{C}$ around the reference temperature $T_{\mathrm{cab,\,ref}} = 21\,°\mathrm{C}$ is an acceptable deviation [LLDL11, VA15]. Therefore, the acceptable temperature range is set to $\Delta T_{\mathrm{cab,\,acc}} = [20\,°\mathrm{C}, 22\,°\mathrm{C}]$. The ambient humidity, the parameters of the vehicle temperature model, and further parameters regarding the HVAC system are given in Table A.2 in the Appendix.

The prototypical implementation and simulative validation are realized in MATLAB. The simulation includes representations of the APN components as load profiles shown in Figure 5.2 and Figure 5.3 and more detailed models, such as the HVAC system model. For the auction-based mechanism, the individual PPFs are separately defined for each component and aggregated in a central function to determine the MCP. The individual PPFs and the central aggregation refer to the mathematical descriptions in Section 4.1. The primal objective is to determine the MCP in (4.10) for the next time step k in order to balance the power consumption in the APN as depicted by the flow chart in Figure 4.6. The MCP holds for $\Delta T = 200\,\mathrm{ms}$. During the time intervals k, the power consumption and supply are regarded as constant in the simulation. Hence, the simulative evaluation focuses on the power balancing mechanism and neglects the voltage and current dynamics in the range of microseconds to milliseconds, which are not within the scope of the thesis.

5.1 Market Price Adaptation with Multiple Voltage Levels

This section demonstrates the general working principle of auction-based power management and its market price adaptation for power balancing. Furthermore, the improved generalization in terms of multiple voltage levels is highlighted in the simulation.

Table 5.1: Parameters of the simulation scenarios with an emphasis on the relevant PPF variables [SGH24].

Description	Variable	Value
Boosting power of the EM	$P_{\text{set, boost}}$	−4750 W
Generation power of the EM	P_{gen}	2000 W
Recuperation power of the EM	$P_{\text{set, rec}}$	6250 W
SOC 48V battery	$\text{SOC}_{48\text{V}}$	0.6
SOH 48V battery	$\text{SOH}_{48\text{V}}$	0.8
Lower charging temperature	$T_{\text{c, min}}$	0 °C
Upper charging temperature	$T_{\text{c, max}}$	45 °C
Lower discharging temperature	$T_{\text{dis, min}}$	20 °C
Upper discharging temperature	$T_{\text{dis, max}}$	60 °C
Temperature 48V battery	T_{Batt}	20 °C
Maximum charging power 48V battery	$P_{\text{c, max}}$	−2000 W
Maximum discharging power 48V battery	$P_{\text{dis, max}}$	4000 W
SOC 12V battery	$\text{SOC}_{12\text{V}}$	0.6
SOH 12V battery	$\text{SOH}_{12\text{V}}$	1
Temperature 12V battery	T_{Batt}	20 °C
Maximum charging power 12V battery	$P_{\text{c, max}}$	−1000 W
Maximum discharging power 12V battery	$P_{\text{dis, max}}$	1500 W
Ambient temperature	T_{amb}	32 °C
Desired temperature in the cabin	$T_{\text{cab, ref}}$	21 °C
Acceptable temperature range	$\Delta T_{\text{cab, acc}}$	[20 °C, 22 °C]
Maximum power of the HVAC	$P_{\text{HVAC, max}}$	3500 W
Normal power of the HVAC	$P_{\text{HVAC, norm}}$	2500 W
Minimum power of the HVAC	$P_{\text{HVAC, min}}$	0 W
Blower level 5	$P_{\text{Blower, 5}}$	1000 W
Blower level 4	$P_{\text{Blower, 4}}$	800 W
Blower level 3	$P_{\text{Blower, 3}}$	600 W
Blower level 2	$P_{\text{Blower, 2}}$	400 W
Blower level 1	$P_{\text{Blower, 1}}$	200 W
Auction interval	ΔT	200 ms
Prediction horizon	N_p	4 s
Control horizon	N_c	4 s

5.1.1 Results

Figure 5.5 depicts the results for the first scenario and the activated auction-based power management. In Figure 5.5a and in Figure 5.5b, the power consumptions in the 12V level and in the 48V level are given, respectively. For the battery storages, Battery 12V and Battery 48V, a positive power represents charging, and a negative power displays the discharging of the respective battery. Simultaneously, the EM power is positive when power is consumed in boost mode and negative when power is generated in adaptive generator or recuperation mode. Positive power graphs depict the power consumption of the various components. As a result of the activated power management, the sum of power, aggregating all consumption and supply in both voltage levels, is zero for each interval k with an interval length of $\Delta T = 200\,\text{ms}$.

The green graph visualizes the market price signal in Figure 5.5c. The price depicts the working principle of the auction mechanism while the region highlighted with the blue stripes represents the normal market price interval $\Delta p_{\text{norm}} = (2, 6]$. For the exemplary scenario, in the time interval in which the EM is boosting the driving operation, the normal price interval $\Delta p_{\text{norm}} = (2, 6]$ is exceeded. In this phase of exaggerated consumption due to the high boosting activity from $k = 1.5\,\text{s}$ to $k = 3.5\,\text{s}$, the market price enters the price interval $\Delta p_{\text{red}} = (6, 8]$. Hence, the power balance is adapted by the market price signal and the according load adaptations. As described in Section 4.1 and in Table 4.1, the comfort load consumption is reduced in phases of limited power availability for a market price $p > p_{\text{red}}$.

The HVAC system, as the main power consumer with the greatest flexibility, adapts its power consumption according to the PPF and the current market price p_k. The resulting power profile is visualized in Figure 5.5d. Due to the high ambient temperature $T_{\text{amb}} = 32\,°\text{C}$ and the radiation by the sun, the HVAC activity causes a mean power consumption of about 1200 W. During the boosting phase, the auction mechanism limits the power consumption so that the HVAC slightly decreases the operation. Consequently, the cabin temperature T_{cab} in Figure 5.5e rises slightly. Since the power reduction is relatively small, the deviation from the desired cabin temperature $T_{\text{cab, ref}} = 21\,°\text{C}$ is below $|\Delta T_{\text{cab, ref}}| = 0.01\,°\text{C}$. Afterward, the HVAC slightly increases its power consumption in order to restore the reference cabin temperature $T_{\text{cab, ref}} = 21\,°\text{C}$. Another part of the power adaptation process is realized by the supplying infrastructure, including the EM and the Battery 48V in Figure 5.5b and the Battery 12V in Figure 5.5a.

During the phases in which the EM is in adaptive generator mode, the market price is about $p = 4$, indicating the normal operation interval. In these phases, the components and systems operate in their normal ranges. As an example, the HVAC system controls its power consumption such that the desired cabin temperature $T_{\text{cab, ref}} = 21\,°\text{C}$ is achieved. Due to the vehicle's energy architecture depicted in Figure 5.1, the EM in the 48V level is mainly responsible for the power supply of both voltage levels. Thus, the 12V level is commonly supported through the DC/DC-converter (PE) by the

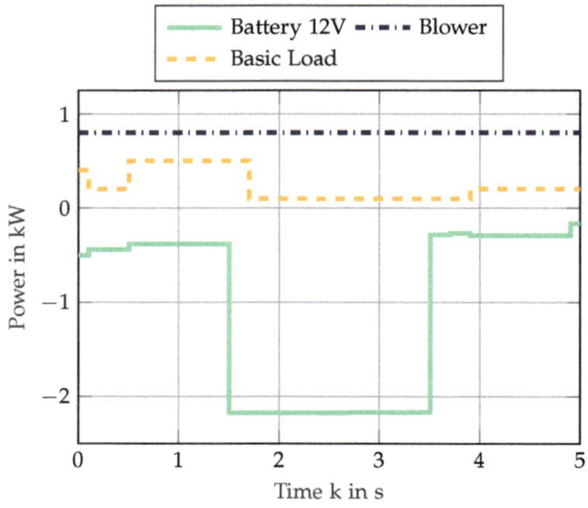

(a) Power consumption in the 12V level.

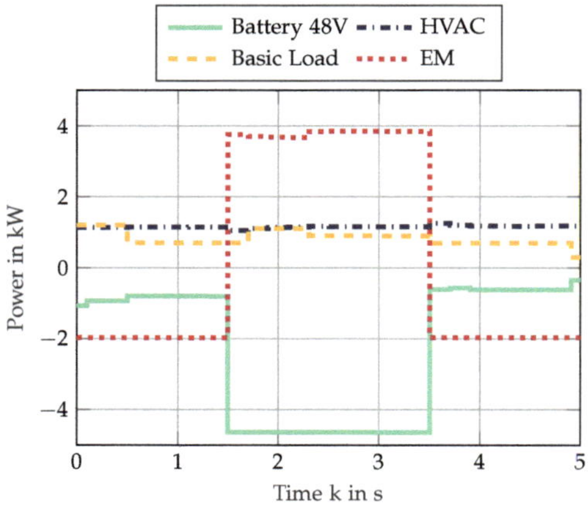

(b) Power consumption in the 48V level.

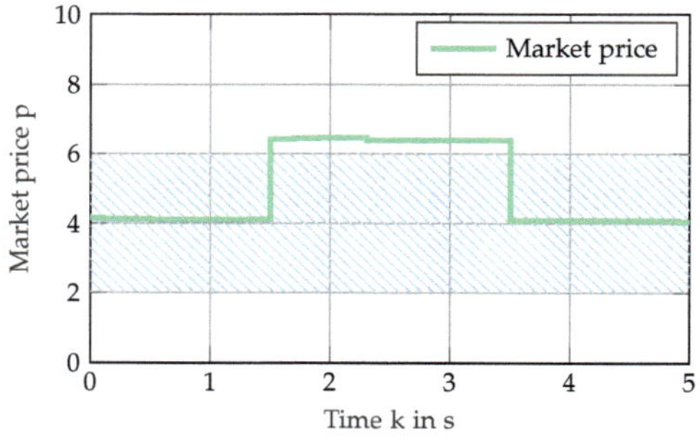

(c) Market price.

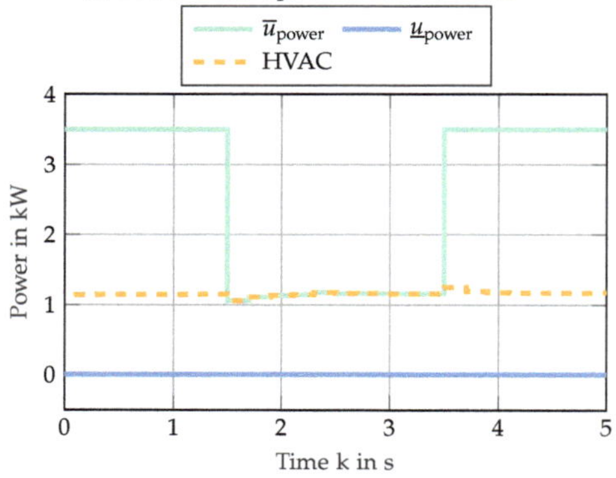

(d) Power consumption of the HVAC system.

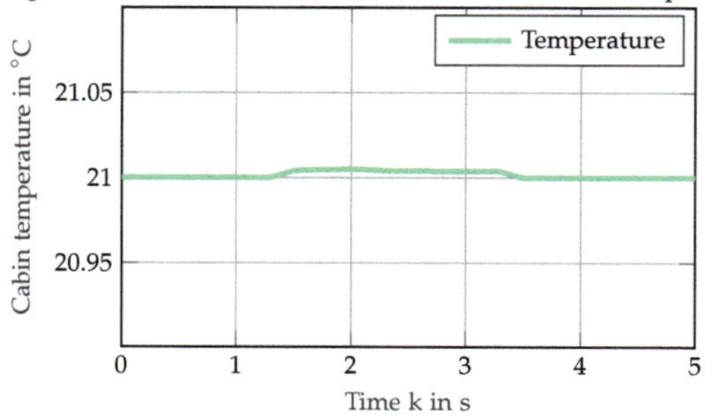

(e) Cabin temperature.

Figure 5.5: Simulation results from the first scenario with the EM in boost mode. The blue stripes in Figure 5.5c mark the normal market price interval $\Delta p_{\text{norm}} = (2, 6]$. Figure 5.5d shows the fulfillment of the power constraints by the MPC resulting in a small temperature deviation in Figure 5.5e.

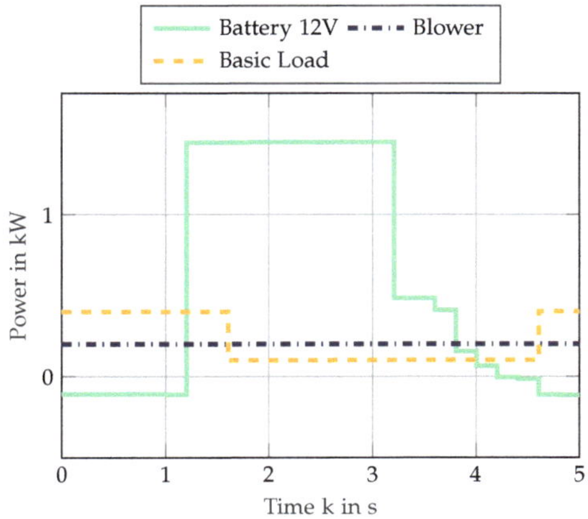

(a) Power consumption in the 12V level.

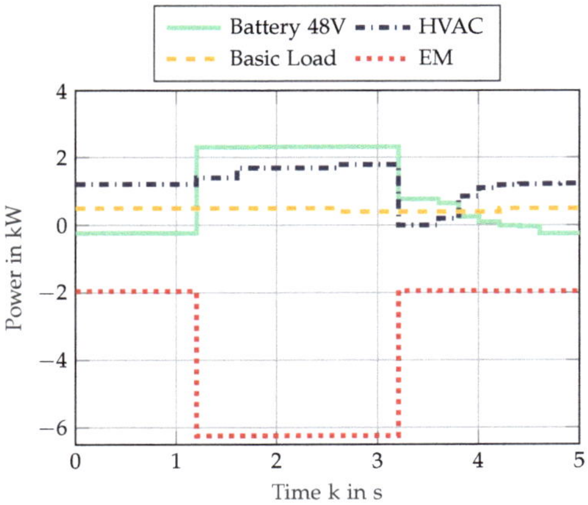

(b) Power consumption in the 48V level.

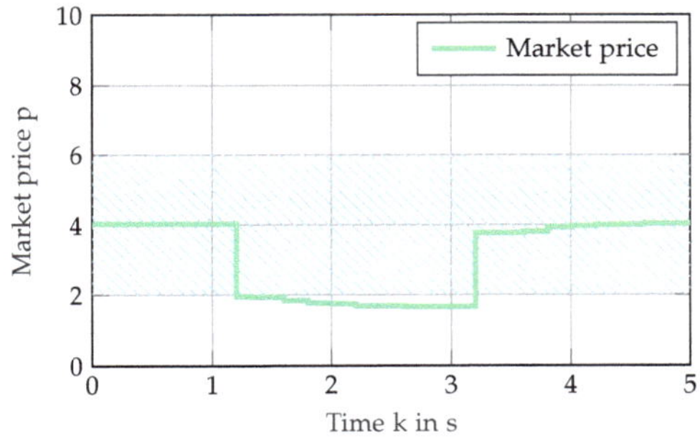

(c) Market price.

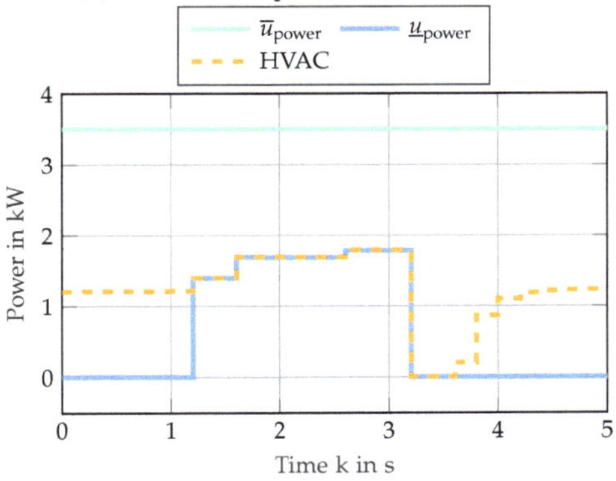

(d) Power consumption of the HVAC system.

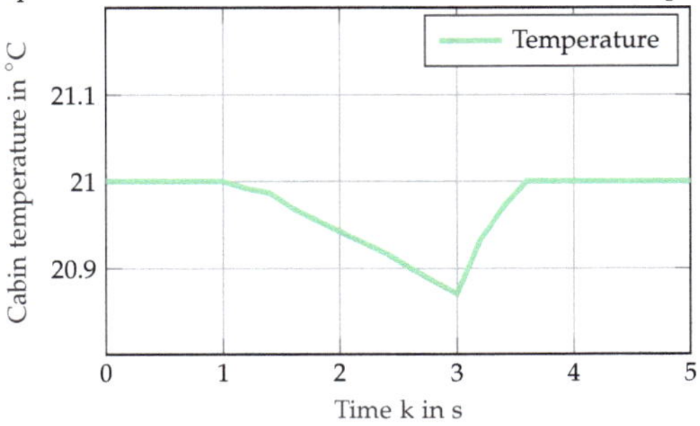

(e) Cabin temperature.

Figure 5.6: Simulation results from the second scenario with the EM in recuperation mode. The blue stripes in Figure 5.5c mark the normal market price interval $\Delta p_{\text{norm}} = (2, 6]$. Caused by the power constraint $\overline{u}_{\text{power}}$ in Figure 5.6d, the cabin temperature T_{cab} deviates from the reference temperature $T_{\text{cab, ref}} = 21\,°\text{C}$.

48V level. Furthermore, the batteries contribute to the power supply in dependence on their SOC as described by the PPF in (4.23).

In the boosting phase, the Battery 12V is highly exploited, as depicted in Figure 5.5a, and the DC/DC-converter transfers the power to the 48V level to support the boosting operation. The Battery 48V is limited in its power capabilities by the SOH $= 0.8$ and the maximum discharging power $P_{\text{dis, max}}$. Since the overall power supply is not sufficient, the EM itself is forced to reduce the boosting from $P_{\text{set}} = -4750\,\text{W}$ to about $-3750\,\text{W}$ in the first second of boosting. Subsequently, the consumption through boosting is increased to about $-3900\,\text{W}$ but still below the desired power $P_{\text{set}} = -4750\,\text{W}$ as shown in Figure 5.5b.

In the second scenario depicted in Figure 5.6, the market price leaves the interval Δp_{norm} caused by a surplus of power through recuperation. Again, the power balance is adapted by the respective market price development. In this case, the consumption of comfort loads is increased due to the elevated power availability, for $p < p_{\text{norm}}$.

Furthermore, a high amount of power is transferred via the DC/DC-converter to the 12V level, supplying the consumers while additionally charging the Battery 12V. With the increased consumption of the HVAC system and the maximum charging power for the Battery 48V, the adaptation caused by the auction mechanism facilitates the full usage of the recuperation power $P_{\text{rec}} = 6250\,\text{W}$ as listed in Table 5.1.

Due to the increased power availability, the HVAC system is asked to consume additional power, leading to a higher cooling action and thereby a decreasing cabin temperature T_{cab} as depicted in Figure 5.6e. Caused by the amount of energy available in the power network, the temperature deviation in the cabin grows to about $|\Delta T_{\text{cab, ref}}| = 0.15\,°\text{C}$ until the recuperation phase ends. Consequently, the HVAC system adapts its power consumption afterward to restore the desired cabin temperature $T_{\text{cab, ref}} = 21\,°\text{C}$ as demonstrated in Figure 5.6d. As a result, the HVAC system saves energy after the recuperation phase, effectively shifting its power consumption.

To show the approach's generalization with regard to multiple voltage levels, the two simulation scenarios are adapted accordingly. Keeping the fundamental scenarios equivalent to the previous description, the PE limitations $P_{\text{PE, lim}}$ are changed from $P^{\text{48-to-12}}_{\text{PE, lim}} = 2500\,\text{W}$ to $\hat{P}^{\text{48-to-12}}_{\text{PE, lim}} = 1500\,\text{W}$ and from $P^{\text{12-to-48}}_{\text{PE, lim}} = 2000\,\text{W}$ to $\hat{P}^{\text{12-to-48}}_{\text{PE, lim}} = 1000\,\text{W}$. The previous and the new PE limitations are listed in Table 5.2. As a result, the balancing capabilities between the connected voltage levels are constrained.

In Figure 5.7, the results for the first scenario with the EM in boost mode are depicted. The derived market price signals in Figure 5.7c strongly differ from Figure 5.5c since the market price distinguishes for the 12V and the 48V level. The global market price in green is comparable to the previous simulation results. However, the market prices show differences between the two voltage levels due to the limited power transfer.

The explanation is provided in Figure 5.7d, which visualizes the power transfer between the voltage levels. With the yellow graph, the actual power transfer toward the

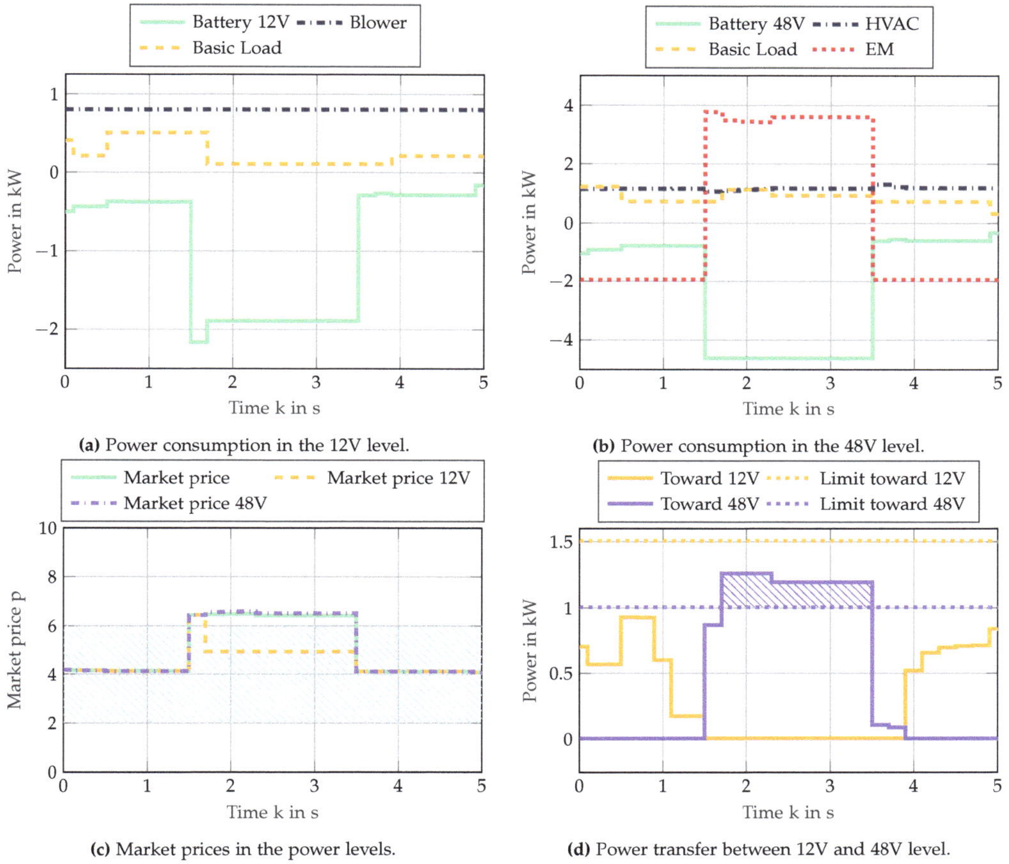

(a) Power consumption in the 12V level.

(b) Power consumption in the 48V level.

(c) Market prices in the power levels.

(d) Power transfer between 12V and 48V level.

Figure 5.7: Simulation results from the first scenario with the EM in boost mode and the lower PE limitations $P_{\mathrm{PE,lim}}$ in Table 5.2. The purple stripes depict the power transfer cut off caused by the power limit toward the 48V level.

12V level $P^{\text{48-to-12}}$ is depicted while the yellow dashed graph marks the current PE limitation $\hat{P}^{\text{48-to-12}}_{\mathrm{PE,lim}}$. This is the expected direction of power transfer, as mentioned before. The purple graph highlights the power transfer in the opposite direction $P^{\text{12-to-48}}$ toward the 48V level. With the purple dashed graph, the PE constraint $\hat{P}^{\text{12-to-48}}_{\mathrm{PE,lim}}$ for this power transfer direction is given.

Due to the high power demand in the boosting phase, the Battery 12V supports the 48V level via the DC/DC-converter. However, the actual power transfer $P^{\text{12-to-48}}$ is limited by the PE limitation $\hat{P}^{\text{12-to-48}}_{\mathrm{PE,lim}}$ in this scenario. The power-transfer limitation is highlighted with the purple pattern. As a consequence, the 48V level has to restrict power consumption further, leading to an increasing market price compared to the standard price in green. The individual market price of the 48V level is visualized with the dashed purple graph in Figure 5.7c. The high basic load in the 12V level,

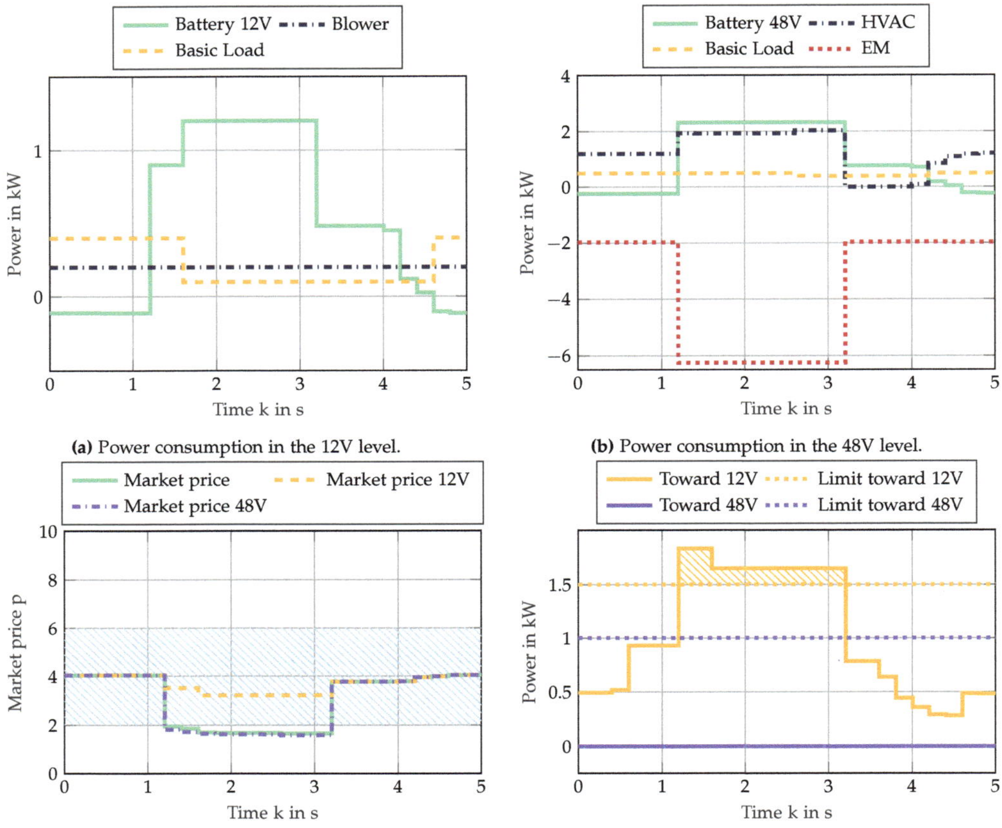

(a) Power consumption in the 12V level.

(b) Power consumption in the 48V level.

(c) Market prices in the power levels.

(d) Power transfer between 12V and 48V level.

Figure 5.8: Simulation results from the second scenario with the EM in recuperation mode and the lower PE limitations $P_{\mathrm{PE,lim}}$ in Table 5.2. The yellow stripes depict the power-transfer cut off caused by the power limit toward the 12V level.

from $k = 1.5\,\mathrm{s}$ to $1.8\,\mathrm{s}$, causes an increased market price in the 12V level, which is given by the yellow dashed graph. After the decrease of the basic load, the market price is recovering to the normal price interval Δp_{norm} and below the previous global market price in green.

In the 48V level, the EM and the HVAC system are forced to compensate for the missing support from the 12V level if compared to the previous simulation. The differences are given in the comparison of Figure 5.5b and Figure 5.7b, visualizing the power consumption and supply in the two simulations.

During the period of regenerative braking in the second scenario in Figure 5.8, the power transfer toward the 12V level is limited. Thus, less of the regenerated electric power can be transferred to charge the Battery 12V. Therefore, the HVAC system

Table 5.2: PE limitations for the simulative evaluation of the auction mechanism for multiple voltage levels.

Description	Variable	Value
Old PE limitation from 48V to 12V	$P_{\mathrm{PE,lim}}^{\text{48-to-12}}$	2500 W
Old PE limitation from 12V to 48V	$P_{\mathrm{PE,lim}}^{\text{12-to-48}}$	2000 W
Reduced PE limitation from 48V to 12V	$\hat{P}_{\mathrm{PE,lim}}^{\text{48-to-12}}$	1500 W
Reduced PE limitation from 12V to 48V	$\hat{P}_{\mathrm{PE,lim}}^{\text{12-to-48}}$	1000 W
PE limitation from 48V to 12V during DC/DC-converter breakdown	$\tilde{P}_{\mathrm{PE,lim}}^{\text{48-to-12}}$	0 W
PE limitation from 12V to 48V during DC/DC-converter breakdown	$\tilde{P}_{\mathrm{PE,lim}}^{\text{12-to-48}}$	0 W

has to compensate for the PE limitation $\hat{P}_{\mathrm{PE,lim}}^{\text{48-to-12}}$, leading to a higher consumption. The yellow pattern in Figure 5.8d highlights the power-transfer limitation toward the 12V level. In Figure 5.8b, the increased consumption in comparison to the previous simulation in Figure 5.6b is depicted. After the recuperation phase, the HVAC reduces the cooling activity to zero, restoring the reference cabin temperature.

Due to the reduced consumption of the HVAC system following the recuperation phase, the Battery 48V is charged for a more extended period, visualized in the green graph in Figure 5.8b. Again, according to the previous simulation, the changing PE limitation leads to deviating market prices for the two voltage levels in Figure 5.8c. Since the power transfer to the 12V level is constrained, the market price for the 48V level in purple decreases, causing the described behavior of the HVAC. On the other hand, the market price 12V in yellow stays in the normal price interval, indicating that the 12V level is not fully involved in the consumption of the additionally available power.

5.1.2 Discussion

The simulation results demonstrate the effectiveness of the auction-based automotive power management and the quality of power balancing. In all situations, a reliable power balance between the respective suppliers and consumers in the APN is achieved, which again is the foundation for successful voltage stabilization. In the adaption process leading to the power balance, all suppliers and consumers take part with reference to their PPF. The necessary adaptations are divided suitably among the APN components. In case of limited power availability in the APN, the proposed auction mechanism reduces power consumption and increases power supply simultaneously. Contrarily, in case of excessive power availability, the mechanism requests further demand and limits the supply. Hence, the proposed auction design represents a capable and holistic procedure, including supply and demand, to reach a balanced power.

The simulation scenarios comprise various modes of the EM, leading to strongly varying power demands and supplies. Hence, the scenarios highlight the relevant situations in which automotive power management is required to balance the power and facilitate a stable and reliable supply. Consequently, the simulations are suitable to demonstrate the capability of the proposed auction-based power management approach. Nevertheless, these scenarios represent rare situations in standard vehicle operations.

The simulations with the power transfer limitations $P_{\mathrm{PE,lim}}$ in Figure 5.7 and in Figure 5.8 point out the improved generalization as described in Definition 2.8. Compared to the concept in [Gra04a], the novel auction-based power management approach handles different APN architecture in terms of multiple voltage levels. With the individual market prices, the procedure adjusts the global MCP and achieves individual power balances within the voltage levels and in compliance with the prevailing power transfer limitations.

The proposed power management algorithm is one shell in the overall APN management system in Figure 2.2. As depicted in Figure 1.1, the approach emphasizes the auxiliary or comfort loads. In contrast, the safety-relevant loads are mainly controlled by the EMS coordinating the driving operation [GS17, HWK+17]. In the simulation, this aspect is considered by basic load profiles, which represent the safety-relevant components and do not take part in the auction. Since the load coordination is mainly limited to comfort loads, achieving a reliable power supply in critical situations depends on a solid power and energy infrastructure in the APN. Thus, in case of severe system failures, the hardware design has to provide enough power to supply all safety-relevant loads in a particular driving situation while the proposed auction mechanism shuts down all comfort loads.

5.2 Market Price Adaptation During Critical Vehicle States

During safety-relevant driving maneuvers, the safe vehicle state depends on a reliable power supply. Safety-relevant components, such as the steering motor or the electric suspension, provide vehicle driving stability. At the same time, these components cause high power consumption and, in particular, power peaks with an amplitude of several kW. Therefore, the following scenario demonstrates the effectiveness of the proposed power balancing in critical situations.

5.2.1 Results

The critical situation builds upon the second scenario, in which the EM is in recuperation mode. Furthermore, the scenario comprises adjustments in the power-transfer capabilities between the two voltage levels, as shown in the previous section. However, in the prevailing scenario, there is a complete DC/DC-converter breakdown, resulting in $\tilde{P}^{48\text{-to-}12}_{\mathrm{PE,lim}} = \tilde{P}^{12\text{-to-}48}_{\mathrm{PE,lim}} = 0\,\mathrm{W}$ in both directions. Figure 5.9 shows further circumstances that are changed compared to the standard recuperation scenario. Since there is only a required power transfer from the 48V level toward the 12V level, Figure 5.10d neglects the power transfer from the 12V level toward the 48V level.

The new load profile of the basic load 12V in Figure 5.9a marks a great power increase compared to the previous load profile in Figure 5.3a. The load profile comprises the power consumption of safety-relevant loads not directly affected by the auction-based power balancing. The maximum power is reached at about $k = 1.5\,\mathrm{s}$ with $2\,\mathrm{kW}$. Additionally, the battery 12V provides a divergent behavior since the SOC is changed from $\mathrm{SOC}_{12\mathrm{V}} = 0.6$ to $\mathrm{SOC}_{12\mathrm{V}} = 0.3$. Consequently, the PPF in Figure 5.9b shifts to higher prices as highlighted by the dark blue graph for the PPF with $\mathrm{SOC}_{12\mathrm{V}} = 0.3$.

The DC/DC-converter breakdown, resulting in $\tilde{P}^{48\text{-to-}12}_{\mathrm{PE,lim}} = \tilde{P}^{12\text{-to-}48}_{\mathrm{PE,lim}} = 0\,\mathrm{W}$, causes two separate power levels. Since there is no functional connection in terms of power transfer, each voltage level, 12V and 48V, has its price as depicted in Figure 5.10c. During the scenario, the market price 12V and the market price 48V differ. While the market price 12V is in the price ranges Δp_{red} and Δp_{crit}, the market price 48V is mostly low. During the recuperation, the market price 48V remains in the price range Δp_{rec}, indicating a surplus of electric power. However, due to the impeded power transfer toward the 12V level, the electric power has to be consumed within the 48V level.

Because of the undermined power supply and the high market price, the blower in the 12V level reduces its power consumption over the whole scenario. Within the phase of critical power supply ($p \geq 8$) from $k = 1.5\,\mathrm{s}$ to $k = 4.5\,\mathrm{s}$, the blower is switched off in order to ensure the power supply of the safety-relevant components represented by

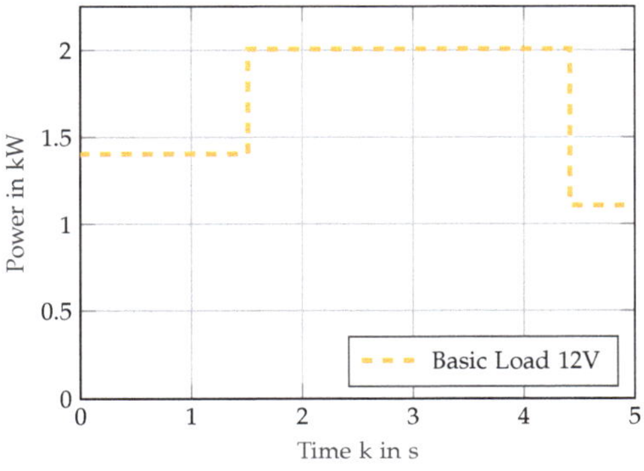

(a) New basic load 12V.

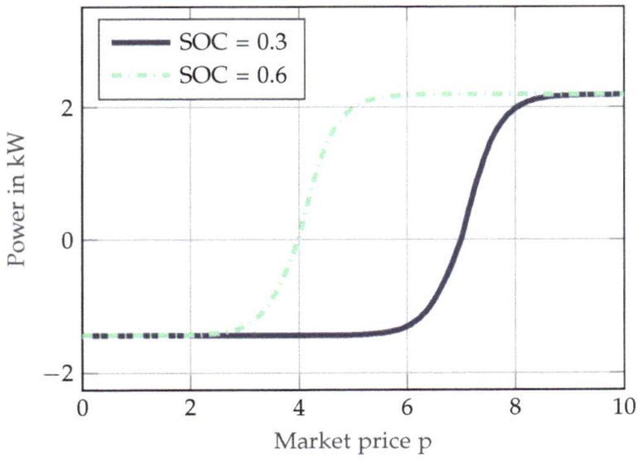

(b) Battery 12V PPFs.

Figure 5.9: New load profile of the basic load 12V and a changing battery 12V SOC with $\mathrm{SOC}_{12\mathrm{V}} = 0.3$ which are applied in the critical scenario with DC/DC-converter breakdown.

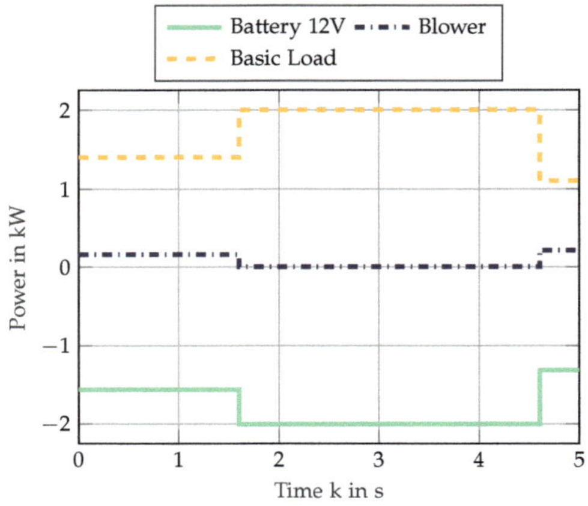

(a) Power consumption in the 12V level.

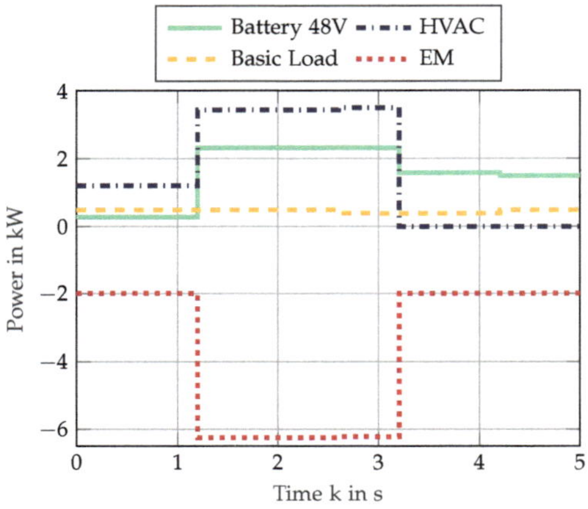

(b) Power consumption in the 48V level.

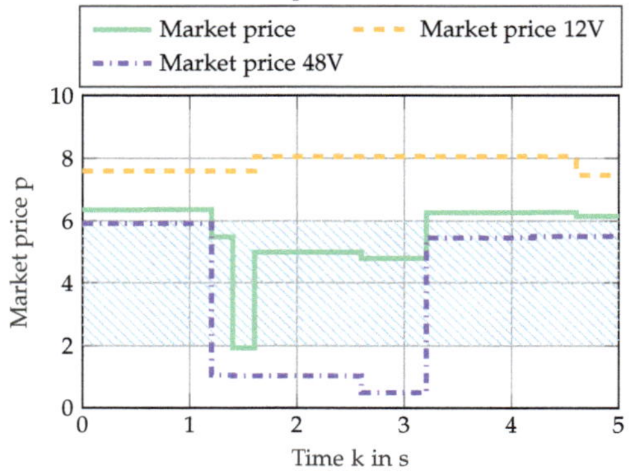

(c) Market prices in the power levels.

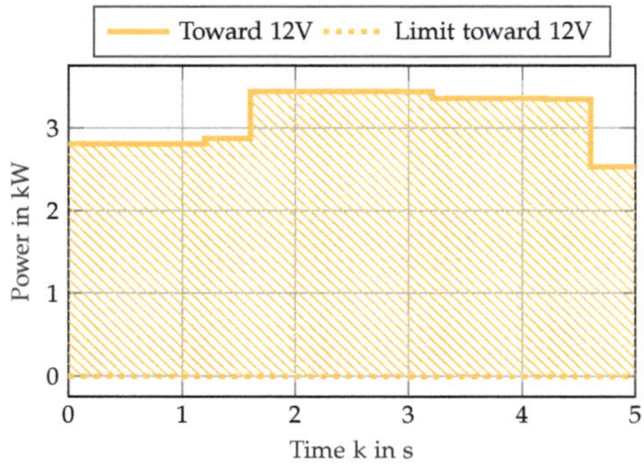

(d) Power transfer from 48V toward 12V level.

Figure 5.10: Simulation results from the second scenario with the EM in recuperation mode and a complete DC/DC-converter breakdown. As a result, the market price in the 12V level is in the critical price range Δp_{crit}, forcing the blower to switch off. Due to the EM in recuperation mode, the market price in the 48V level is low and within Δp_{rec}.

the basic load profile. During this phase, the battery 12V is at the maximum discharge power with about 2 kW. The yellow highlighted area in Figure 5.10d shows the electric power that would be transferred toward the 12V level if the DC/DC-converter worked adequately. In the described situation, the need for power transfer results from the lack of power in the 12V level and the surplus of power through regenerative braking in the 48V level.

In the 48V level, the power surplus during the recuperation phase is compensated by the HVAC system depicted by the dark blue dashed-dotted graph. While the EM provides power with $P_{\text{set, rec}} = -6250\,\text{W}$, the HVAC works at its maximum power consumption $P_{\text{HVAC, max}} = 3500\,\text{W}$. At about $k = 3\,\text{s}$, the market price 48V visualized in Figure 5.10c drops further and forces the EM to reduce the amount of regenerated electric power. After the recuperation phase, the HVAC system is switched off since the temperature in the cabin has moved away from the reference cabin temperature $T_{\text{cab, ref}}$.

5.2.2 Discussion

The presented simulation results demonstrate the power balancing effectiveness even during emergency mode. Since the DC/DC-converter breakdown seriously impairs the electric power supply in the 12V level, the auction-based power balancing approach is strongly demanded. Nevertheless, the critical power supply, which is indicated by the market price range Δp_{crit} and a market price $p \geq 8$, is managed by a consequent power reduction of comfort loads. Consequently, the blower in Figure 5.10a is shut down during the highest power demand. Hence, the simulation results highlight the control of critical situations and assurance of a safe vehicle operation.

From the hardware perspective, the results visualize the importance of the DC/DC-converter connecting the different voltage levels. The electric power transfer in APNs with multiple voltage levels is essential to operate the vehicle and to reduce costs and resources. If the voltage levels are separated, each level has to provide more power capabilities than a connected APN in which the voltage levels support and stabilize each other. In the prevailing APN configuration in Figure 5.1, there are, in particular, two important cases for power transfer. On the one hand, the 12V level does not need a self-reliant electric power generation since it is supported by the 48V level and the EM. On the other hand, the EM is able to regenerate a higher amount of electric power through recuperation because the power surplus can be consumed or stored in the 12V level. These advantages are marked by the green mutual market price, which represents the market price in terms of a functional power transfer and demonstrates the averaging effect of the DC/DC-converter connection.

Regarding the auction algorithm, the DC/DC-converter breakdown and the impeded power transfer between the voltage levels decouple the two APN parts. Thus, the market price determination in Figure 4.6 leads to individual market prices p^m with $m \in \mathbb{V} = \{12\text{V}, 48\text{V}\}$. Without the DC/DC-converter connection, the auction-based power balancing performs two separate auctions.

Another essential aspect in the simulation results is the APN flexibility regarding volatile power consumption and critical component failure. In the presented scenario, the lack of power supply is counterbalanced by the blower, which is switched off. Additionally, the battery is discharged with its maximum discharging power. In summary, flexible loads and storage systems within a power level provide the capabilities to compensate for safety-relevant power shortages. Hence, the interplay of these components and an effective power management ensures the safe vehicle operation.

5.3 Influence of Stepwise Switching Loads

The power balance in the APN is disturbed by various effects. This section shows the simulation results regarding the influence of stepwise switching loads. Due to the deviation between the PPF and the hysteresis function, the switching loads cause a deviation from the calculated power balance.

5.3.1 Results

The stepwise switching loads comprise jumps in the power demand due to the use of different and discrete resistor values. As a result, the cumulative PPF in Figure 4.5 may include two power levels referring to one market price. Furthermore, due to the power steps in the cumulative PPF, the precise calculation of the MCP would be difficult or impossible [Büc08].

Therefore, a continuous PPF in compliance with the Definition 4.1 and a hysteresis mechanism are implemented for the stepwise switching comfort loads actively participating in the auction. However, the hysteresis mechanism, which prevents cyclic switching actions, causes differences between the calculation for the MCP and the actual power consumption or supply. In Figure 5.11a, the blue dashed graph depicts the power consumption according to the blower's PPF in the first scenario with the lower $SOC_{48V} = 0.4$ (see Figure 5.13). Contrarily, the green graph represents the power consumption caused by the blower's hysteresis design, which is visualized in Figure 4.11.

First, the blower holds the previous power level $P_{\text{Blower},4} = 800\,\text{W}$, which the passenger sets. Then the blue dashed graph exceeds the limit for the next power level at about $k = 1.6\,\text{s}$ forcing the hysteresis mechanism to reduce the consumption to the

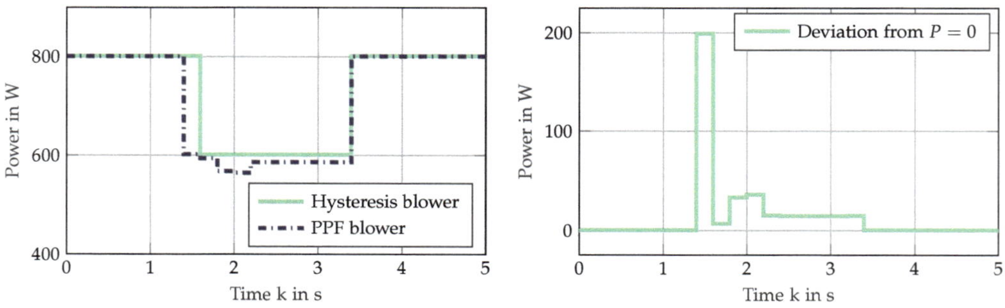

(a) Blower PPF and hysteresis.

(b) Deviation between the blower PPF and the hysteresis.

Figure 5.11: Graphs of the blower's power consumption on the left side and the deviation from the calculated power balance due to the hysteresis design for switching loads as visualized in Figure 4.11.

next lower power level $P_{\text{Blower},3} = 600\,\text{W}$. This hysteresis level is kept until the PPF returns to the original power level of $P_{\text{Blower},4} = 800\,\text{W}$.

The absolute deviation between the PPF and the actual consumption is visualized in Figure 5.11b. The maximal deviation between the PPF and the actual power demand is reached at about $k = 1.5\,\text{s}$ with $\Delta P = 200$ W.

5.3.2 Discussion

A drawback of the stepwise switching components and their hysteresis are the deviations from the calculated power balance depicted in Figure 5.11b. This leads to an unbalanced power in the APN that is mainly compensated by the passive energy storage, for instance, the batteries, or by the underlying PES, for example, by the DC/DC-converter or the EM [RBW+12]. As a result, this aspect influences the battery's exact charging or discharging current or other storage systems.

The short-term power peaks of safety-relevant components, such as the electric suspension, cause a similar effect since they are often unpredictable. Hence, these power peaks interfere with the achieved power balance. To reduce the severeness of this issue, a suitable approach is the adjustment of the interval time ΔT. If the interval length is reduced, the power management would adjust the power balance more frequently to the current situation, resulting in less power deviation.

The proposed mechanism to mitigate the hysteresis effect is described in Section 4.1.4. With the global coordination of the power steps in the respective price ranges, the mechanism ensures that the power steps for switching are equally distributed. Hence, the maximum deviation between the calculated power balance $\text{PPF}_{\text{cum}} = 0$ and the actual supply and demand is kept within one power step of a APN component. So, as an example, the proposed mechanism prevents the blower and the seat heating to locate their power steps in the same market price range.

Another idea is to emphasize the continuously controllable loads and include the stepwise switching loads only in severe situations if all power demand of comfort loads has to be shut down. As a result, the continuously controllable loads, such as the HVAC system, would reduce their power at the beginning of the market price range Δp_{red}. On the other hand, the stepwise switching loads, such as the blower, would adjust their power consumption at the end of the respective market price range.

A general trend that diminishes the drawback and adds further flexibility to the APN is the ongoing dissemination of PEs, such as DC/DC-converter, which improves the control performance of most loads in the APN [Ema05, EWK06]. Consequently, by applying PE, the stepwise switching loads become rare in vehicles.

5.4 Analysis of Individualization via Price-to-Power Functions

As elaborated in Section 4.1.3, the PPFs adapt to the component states and additional information. The facilitated individualization, which is desirable to adapt the auction mechanism to individual component behavior and user preferences, is described in Definition 2.9. This section highlights and discusses the effects of PPF variations through this individualization.

5.4.1 Results

In the following simulations, the fundamental simulation design is similar to the previous description and the parameters given in Table 5.1. However, to point out the influence of individual PPFs, the SOC for the Battery 48V is changed from $SOC_{48V} = 0.6$ to $SOC_{48V} = 0.4$. Since a lower SOC represents a lower charging state, the individual PPF adapts as depicted in Figure 5.12. By the shift to the right side, the PPF with a lower SOC affects a different battery behavior over the price range, resulting in longer periods of charging and shorter periods of discharging in accordance with the market price. The proposed auction-based power management adaptations are visualized in Figure 5.13 for the first scenario and in Figure 5.14 for the second scenario.

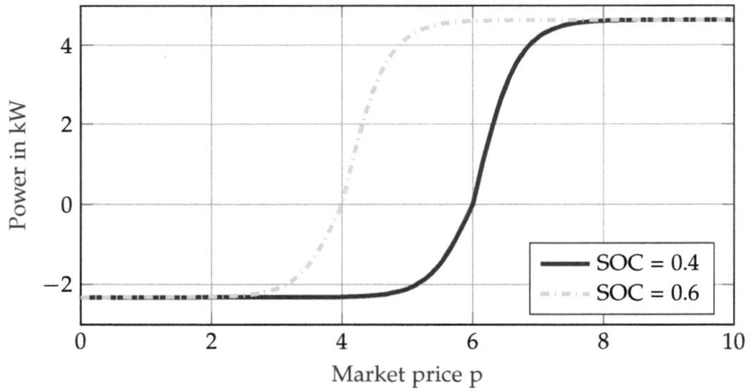

Figure 5.12: Battery 48V PPFs with an $SOC_{48V} = 0.4$ and an $SOC_{48V} = 0.6$.

Figure 5.13c shows the market price development during the first scenario with the EM in boost mode. In comparison to the previous simulation with an $SOC_{48V} = 0.6$ in Figure 5.5c, the market price shifts toward a higher price over the complete scenario.

In the power consumption and supply, the intended battery behavior is observable by the green graph in Figure 5.13b highlighting the charging or discharging of the Battery 48V. Compared to the previous simulations, the battery charges most of the time. The only time interval in which the Battery 48V supplies the 48V level is the

EM boosting phase from $k = 1.5\,\text{s}$ to $k = 3.5\,\text{s}$. Due to the reduced supply, the HVAC system in Figure 5.13d and the EM in Figure 5.13b are forced to limit their power consumption so that the EM boosting is limited to about 3000 W.

The changes in the 12V level are another adaptation aspect displayed in Figure 5.13a. Since the Battery 12V has a higher SOC than the Battery 48V, the Battery 12V provides a higher amount of power to the level 12V and even contributes to the power supply of the 48V level. Hence, the 12V level supports the 48V level via the DC/DC-converter in the prevailing scenario. As a result of the higher market price in the boosting phase and the power transfer to the 48V level, the blower in the 12V level has to be switched to a lower power level $P_{\text{Blower},3} = 600\,\text{W}$ as shown in Figure 5.13a.

Similar to the first scenario, the market price in the second scenario is mostly shifted to a higher price due to the less capable Battery 48V caused by the lower $\text{SOC}_{48\text{V}} = 0.4$.

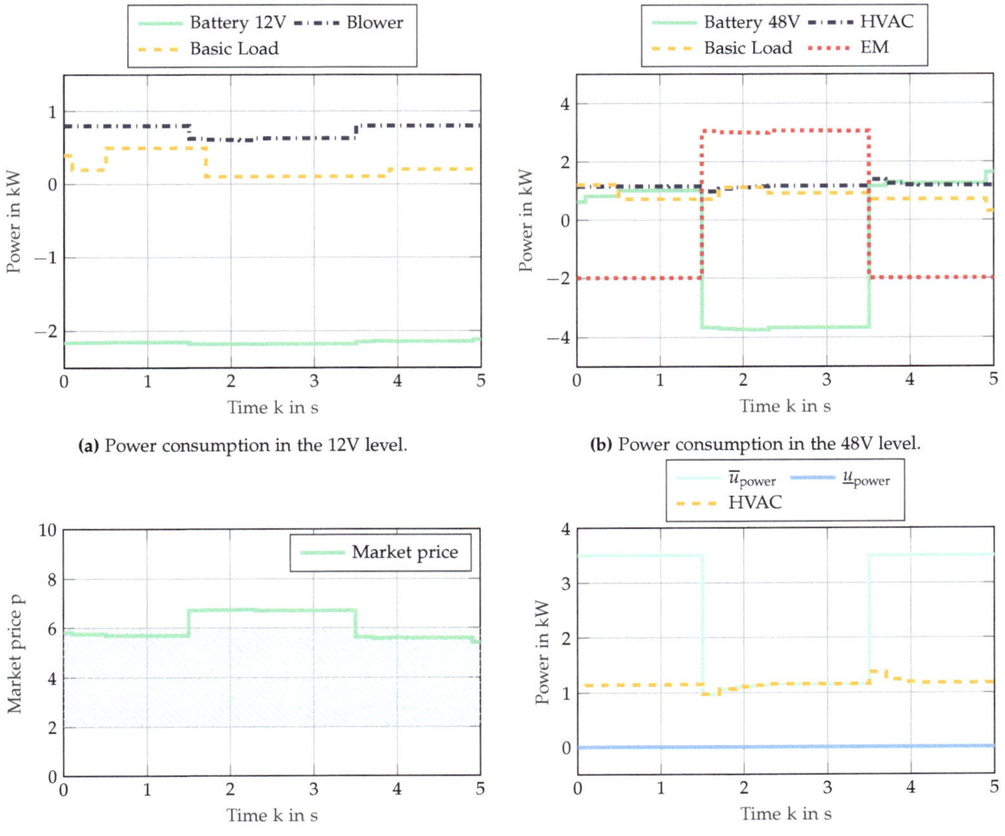

(a) Power consumption in the 12V level.

(b) Power consumption in the 48V level.

(c) Market price.

(d) Power consumption of the HVAC system.

Figure 5.13: Simulation results from the first scenario with the EM in boost mode and an $\text{SOC}_{48\text{V}} = 0.4$ for the battery 48V.

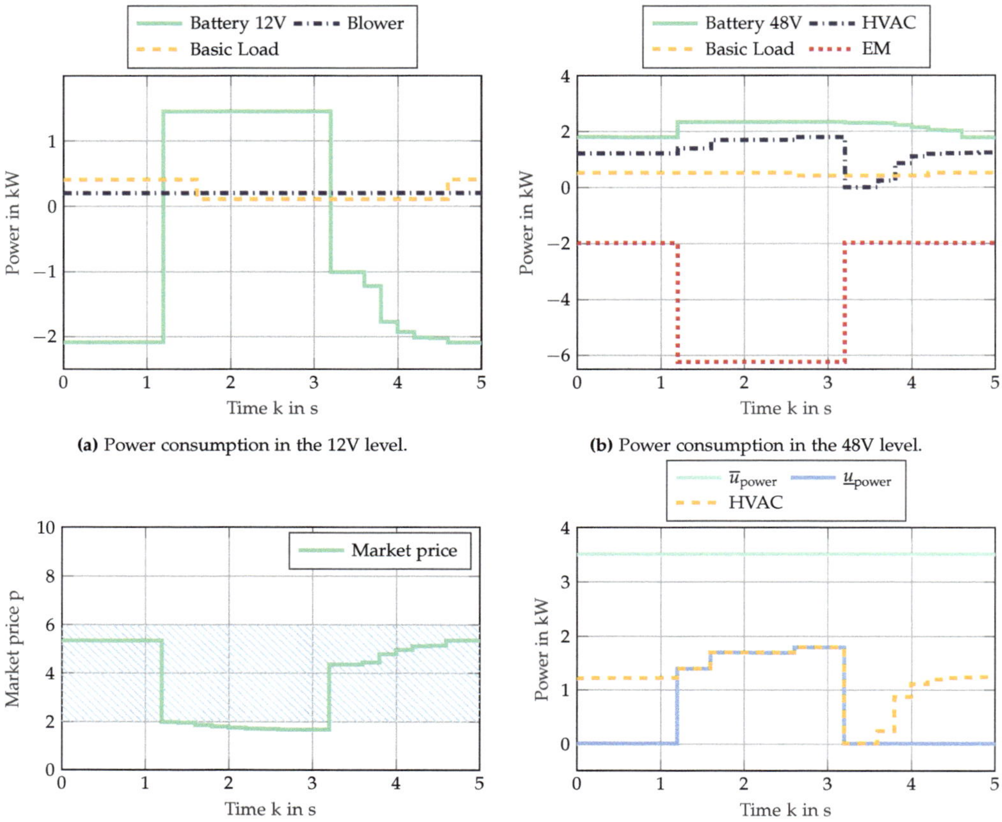

(a) Power consumption in the 12V level.

(b) Power consumption in the 48V level.

(c) Market price.

(d) Power consumption of the HVAC system.

Figure 5.14: Simulation results from the second scenario with the EM in recuperation mode and an $\text{SOC}_{48\text{V}} = 0.4$ for the battery 48V.

The simulation results in Figure 5.14b regarding the component behavior correlate with the according market price in Figure 5.14c. During the whole scenario, the Battery 48V is charging. Nevertheless, the market price leaves the normal price range $\Delta p_{\text{norm}} = (2, 6]$ during the recuperation phase.

5.4.2 Discussion

In cases of a weakened power supply, the market price tends to higher values to acquire more power and, at the same time, limit the power consumption of comfort loads, such as the HVAC system or the blower. With regard to the changed battery SOC, the simulation results change strongly. Since the battery 48V is a less capable power supply, the other components in the APN, supply and demand, have to compensate accordingly. Nevertheless, the auction mechanism works effectively and coordinates the power consumption and supply to achieve a high power balancing quality.

During the recuperation in the second scenario with the battery $SOC_{48V} = 0.4$, the battery 48V is not able to use the complete power for charging. This is due to the limited charging capability caused by the $SOH_{48V} = 0.8$ and by the maximum charging power $P_{c,max}$ in Table 5.1. In this context, the charging and discharging capabilities strongly depend on the battery technology and design [EGLE18, Nea20]. By the applied individualization for the battery PPF in (4.23), these battery characteristics are integrated into the auction process. Hence, the proposed auction-based power management implicitly considers important component characteristics and limitations. This applies for all components independent of when they are added to the APN.

5.5 Analysis of Plug-and-Play Capability

In the following, the plug-and-play integration of new components is analyzed. To demonstrate the desired plug-and-play property, additional loads are integrated into the APN and included in the auction mechanism.

5.5.1 Results

The new components called added load 12V and added load 48V, which are plugged into the existing APN after the HEV's production, are visualized in Figure 5.15. The added load profiles are given for both driving scenarios. For the first scenario with the boosting activity, the additional load profiles are depicted in Figure 5.15a for the 12V level and in Figure 5.15b for the 48V level. Similarly, for the second driving scenario, the additional load profiles are given in Figure 5.15c and Figure 5.15d. Since the loads are integrated after the HEV's configuration and production, the overall APN composition has to provide certain flexibility in terms of connectors and supplying power infrastructure. In this work, the necessary hardware flexibility is assumed to be fulfilled.

The resulting power balance and the consequences regarding the auction-based power management are demonstrated in Figure 5.16 and Figure 5.17. Two additional load profiles for the 12V and the 48V level are included in the prevailing simulation. In comparison to the first simulation, these added components increase the power requirements at both voltage levels. As a result, the market price in Figure 5.16c is slightly shifted to a higher price during the simulation. Hence, the APN copes with the additional load through an increased supply or, if necessary, through a reduction in the comfort load consumption, for example, regarding the HVAC system in Figure 5.16b or the blower in Figure 5.16a.

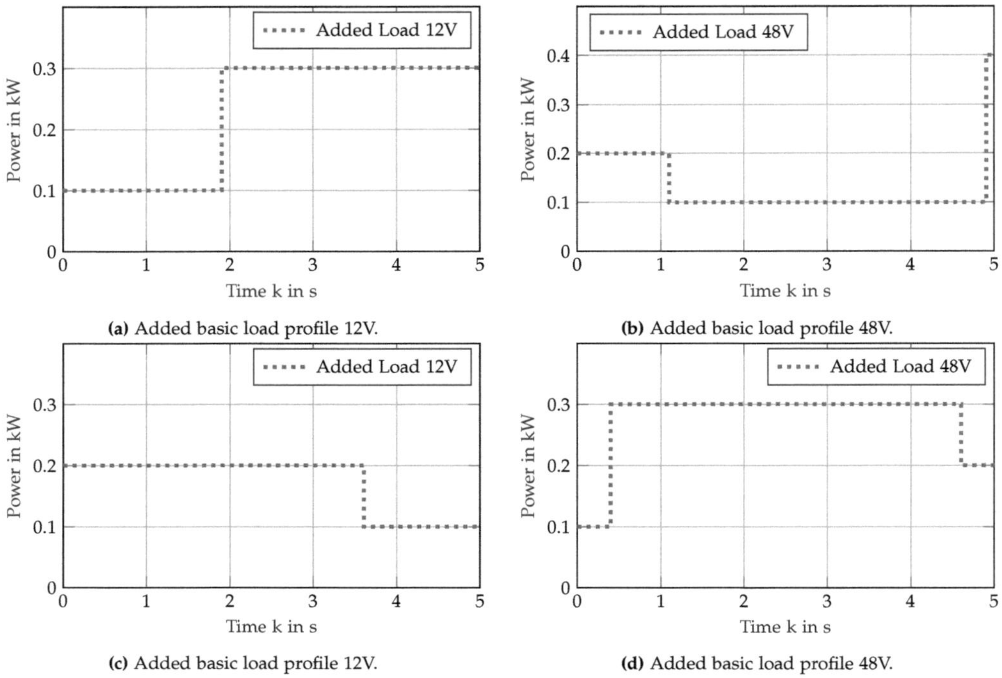

Figure 5.15: The added loads are plugged into the respective voltage level to evaluate the plug-and-play integration property of the proposed auction-based power management approach.

In detail, the blower adjusts its power level during the boosting phase in order to compensate for the additional load in the 12V level. Further adjustments are related to the EM and the HVAC system in the 48V level. Since the overall power consumption in the auction is increased, the EM limits the boosting activity to about 3600 W.

With the elevated market price, the Battery 48V and the Battery 12V enforce their power supply. On the other hand, during the recuperation period in the second scenario displayed in Figure 5.17, the HVAC system limits the additional power consumption due to the overall higher consumption in the power network. In conclusion, the plug-and-play integration in the proposed power management is achieved by adding new components as participants in the auction mechanism. The actual power balancing with all components is then guaranteed through the market price determination in (4.10), which tends to a higher price due to the additional power consumption in the power levels.

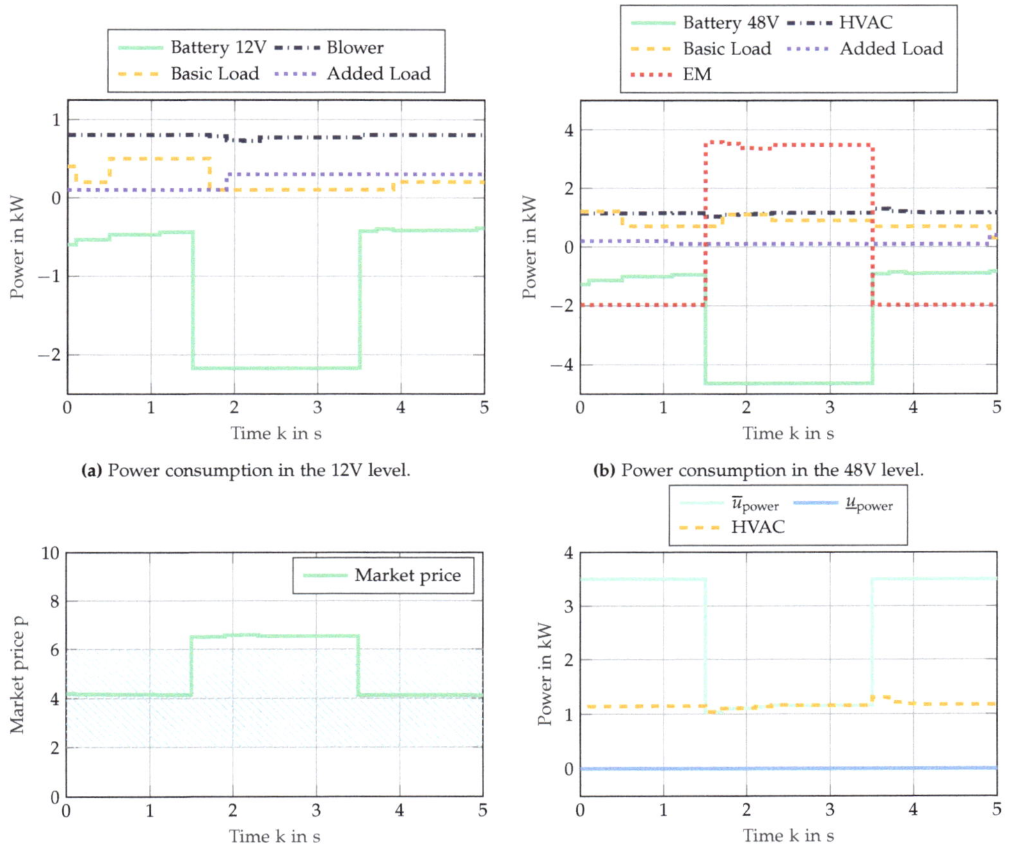

(a) Power consumption in the 12V level.

(b) Power consumption in the 48V level.

(c) Market price.

(d) Power consumption of the HVAC.

Figure 5.16: Simulation results from the first scenario with the EM in boost mode and added basic loads highlighting the plug-and-play integration.

5.5.2 Discussion

Besides the power balancing quality in the APN, the main goal in this thesis is improved flexibility and the achievement of plug-and-play hardware integration. An advantage of the proposed auction-based power management is the possibility of seamless plug-and-play integration of new hardware components by simply adding a PPF to the auction mechanism. This feature includes the integration of further energy infrastructure components, such as energy storages and new loads. Since loads that support the driving operation or have other safety-relevant aspects are not actively involved in the auction, the integration is limited to considering the power balance calculation and MCP determination. Contrarily, controllable comfort loads may play an active role in the adaptation process as they offer the possibility for load reduc-

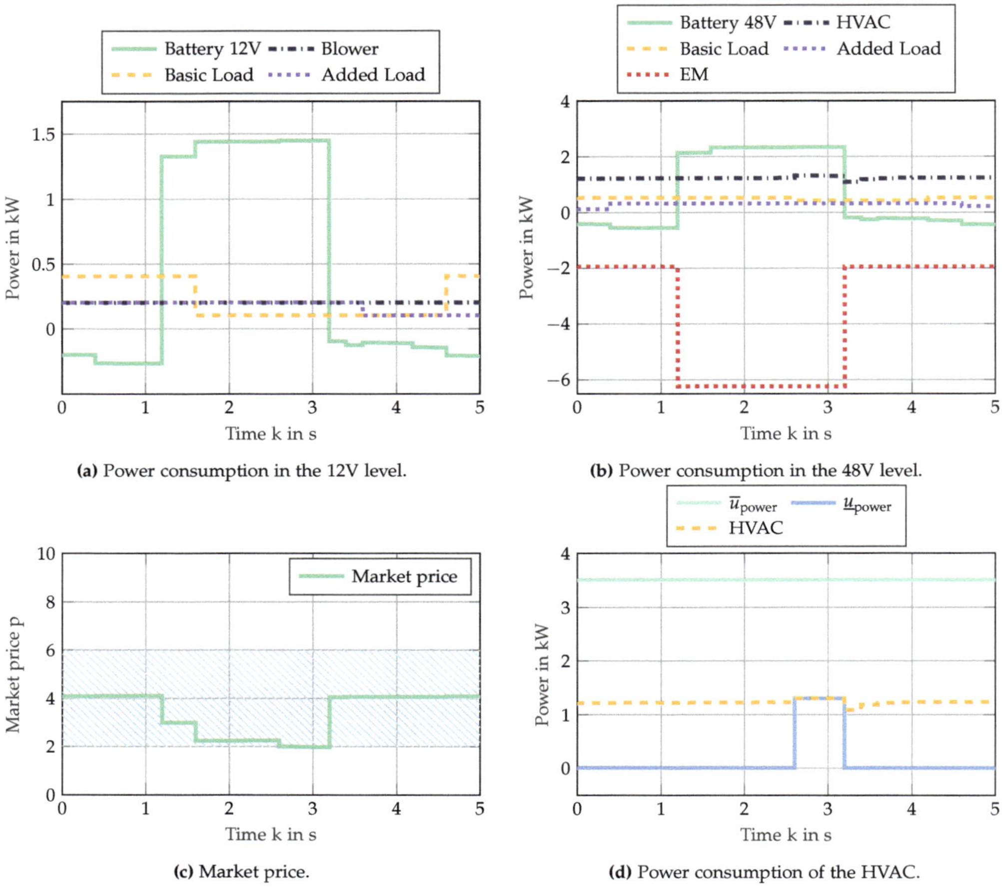

Figure 5.17: Simulation results from the second scenario with the EM in recuperation mode and added basic loads highlighting the plug-and-play integration.

tion like the HVAC system. Consequently, new components are added to the group of sellers or bidders and are taken into account for the market price determination without adapting or changing the overall auction framework.

The plug-and-play integration is effortlessly scalable except for hardware limitations in terms of power supply infrastructure or power and communication connectors. Hence, the necessary communication effort increases, but the computational load within the auctioneer remains constant and is fairly independent of the number of participants. Additionally, the mandatory price ranges in Table 4.1 ensure a reliable auction framework regarding effective power balancing. In summary, the proposed auction-based power management supports a seamless plug-and-play integration of new hardware components and features high scalability.

The communication design plays a significant role in ensuring the actual plug-and-play integration. Since the mechanism includes a central component, namely the auctioneer, the communication schemes have to provide a procedure to integrate the respective hardware components into the auction. This may be done by providing the services in the SOA service distribution as proposed in Section 4.3. In this regard, the described communication design based on the SOA suitably complements the auction-based power management.

5.6 Predictive Extension Exploiting the HVAC Flexibility

The formulated auction-based power management approach in the previous section makes use of the auction mechanism and market price signals to coordinate power balancing measures for the next time interval k. Accordingly, the components in the APN adjust their respective power consumption and power supply behavior based on the PPF and the market price p.

With the predictive extension in Section 4.2, information about the future power consumption and supply is utilized to adapt the load profiles of flexible comfort loads, such as the HVAC system. To adjust the power consumption in an adequate manner, an HVAC model for the description of the temperature dynamics and the required power for cooling or heating is implemented. With the HVAC model and an MPC, the predictive extension provides a trade-off between the precise temperature control, on the one hand, and the compliance with a given power constraint, on the other hand.

5.6.1 Load Shifting and Predictive Power Balance

The following simulation results demonstrate the effectiveness of the predictive extension. In order to allow for a simple comparison with the basic auction-based power management, the simulation scenarios from the previous sections are reused. For the market price signal, the graphs show mostly the price for the basic and the extended approach in direct comparison. Appendix A provides a detailed HVAC model and the according parameters.

Standard Scenarios

Regarding the first standard simulation scenario, Figure 5.18a and Figure 5.18b visualize the power consumption and supply of the involved components in the 12V and in the 48V level, respectively. In Figure 5.5a and Figure 5.5b, the respective power

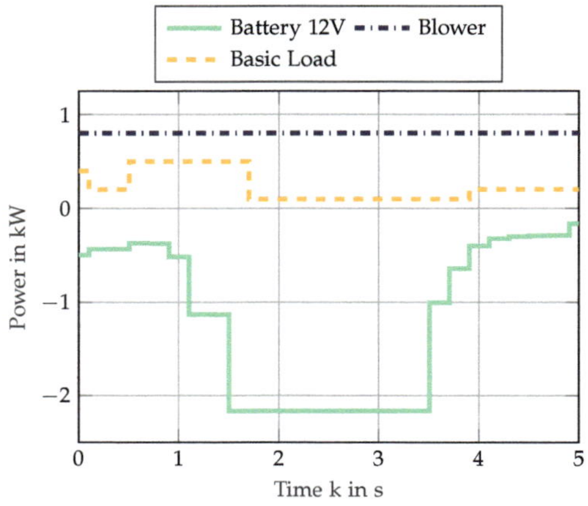

(a) Power consumption in the 12V level.

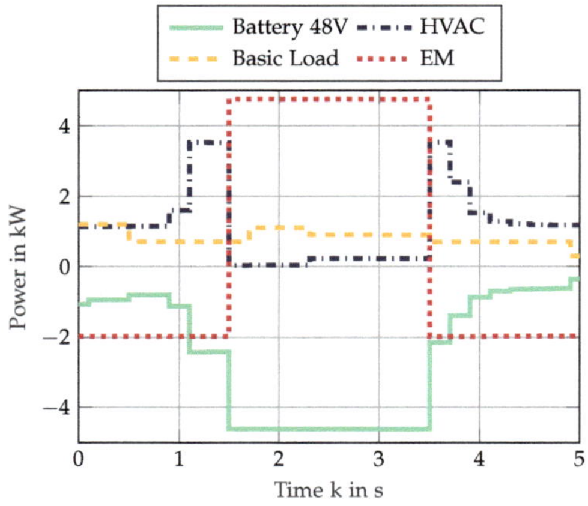

(b) Power consumption in the 48V level.

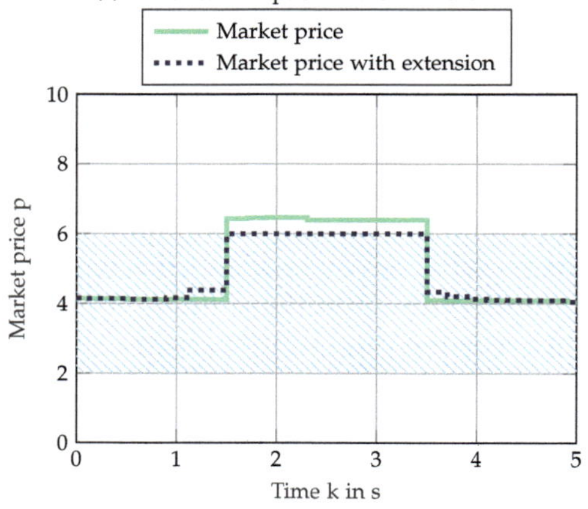

(c) Market prices for the basic and predictive approach.

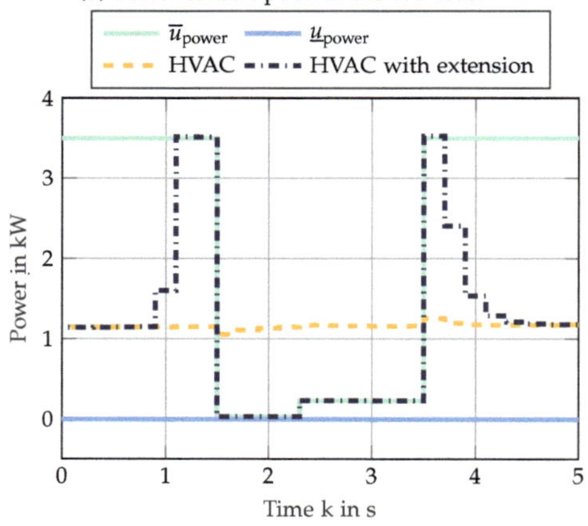

(d) Power consumption of the HVAC system.

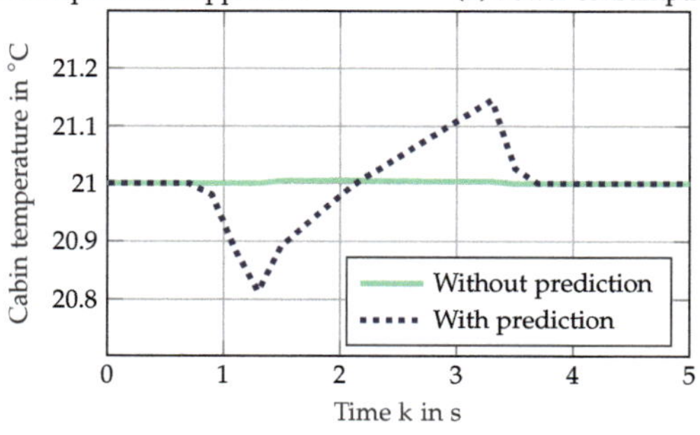

(e) Temperature for the basic and predictive approach.

Figure 5.18: Simulation results from the first scenario with the EM in boost mode and the activated predictive extension shifting the HVAC power consumption.

supply and consumption for the basic approach without the predictive extension is given.

The market price with the predictive extension in the blue dashed graph in Figure 5.18c shows the mechanism's effectiveness. During the whole simulation, the market price stays in the normal price interval $\Delta p_{\text{norm}} = (2, 6]$ while the market price with the basic approach in green exceeds the price threshold $p_{\text{norm}} = 6$ during the boost phase.

The difference from the basic approach is most visible in this extreme situation during boosting activity and in the recuperation period in the second scenario. In both intervals, the market price of the basic approach in green exceeds the borders of the normal price interval $\Delta p_{\text{norm}} = (2, 6]$, resulting in the needed behavior adjustments in the APN to achieve the power balance. Contrarily, the market price of the extended approach given by the blue dashed graph stays within the normal range. Both market prices in direct comparison are depicted in Figure 5.18c.

As a consequence of the active predictive extension, the EM is able to perform the boosting from $k = 1.5\,\text{s}$ to $k = 3.5\,\text{s}$ with the requested electric power $P_{\text{set}} = 4750\,\text{W}$ (see Figure 5.18b). Instead, the HVAC system adapts its power consumption and shifts the load around the period of high power demands in the APN. Figure 5.18d points out the load shifting effort and the precise compliance of the power constraints, which are calculated from the predicted power balance in (4.36). Since the HVAC system comprises various storage capabilities, such as the cabin air or the cooling liquid, the effects on the temperature are within $|\Delta T_{\text{cab, ref}}| = 0.2\,°\text{C}$ as shown in Figure 5.18e. The HVAC control exactly considers the power limits for the upper constraint. The necessary power for the cabin cooling is shifted before the respective period and in parts following the period, keeping the absolute temperature deviation $|\Delta T_{\text{cab, ref}}|$ minimal. In Figure 5.18e, the mechanism and effect are pointed out by the shape of the temperature graph, which shows a precooling in advance to the load shifting interval. After the interval of exceeded power consumption, the HVAC system compensates the previous power savings and restores the desired cabin temperature $T_{\text{cab, ref}} = 21\,°\text{C}$.

Since the load shifting affects the whole APN, the impact is also visible in the power supply and consumption of the batteries. Both batteries, the battery 12V and battery 48V, adapt their behavior based on the overall power balance. The precooling in the shown scenario leads to an increased power supply compared to the basic auction approach in Figure 5.5. In detail, the batteries provide additional power before and following the EM boosting to support the cooling activity of the HVAC system, which is shifted to the sides of the boosting period.

Power Transfer Limitations

In the simulation with PE limitations, power transfer between the two voltage levels, 12V and 48V, is constrained. Thus, the power transfer from the 12V toward the 48V

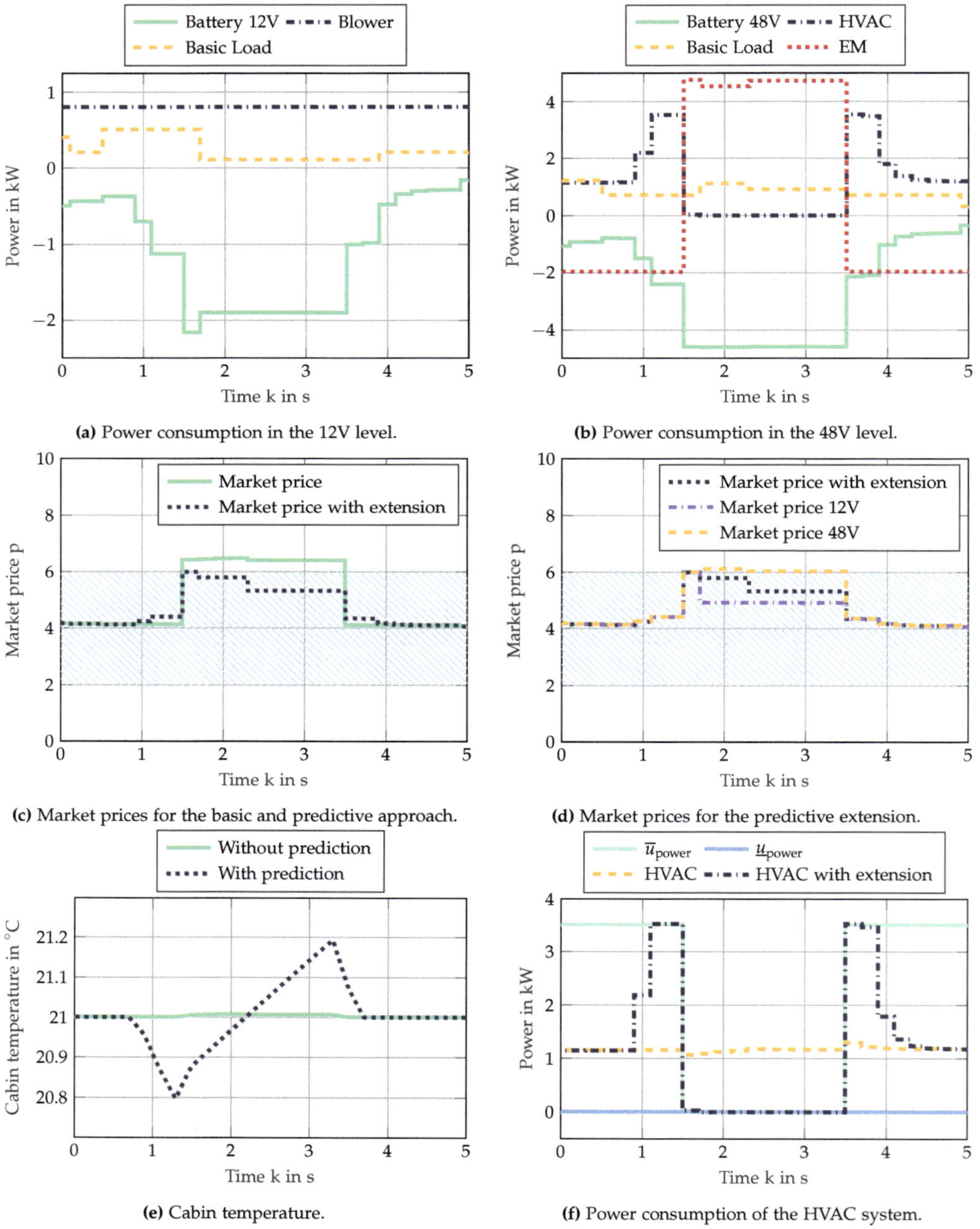

(a) Power consumption in the 12V level.

(b) Power consumption in the 48V level.

(c) Market prices for the basic and predictive approach.

(d) Market prices for the predictive extension.

(e) Cabin temperature.

(f) Power consumption of the HVAC system.

Figure 5.19: Simulation results from the first scenario with the EM in boost mode, the activated predictive extension, and limitations in terms of PE power transfer. The predictive extension keeps the market price mostly within the normal price interval Δp_{norm} marked by the blue stripes.

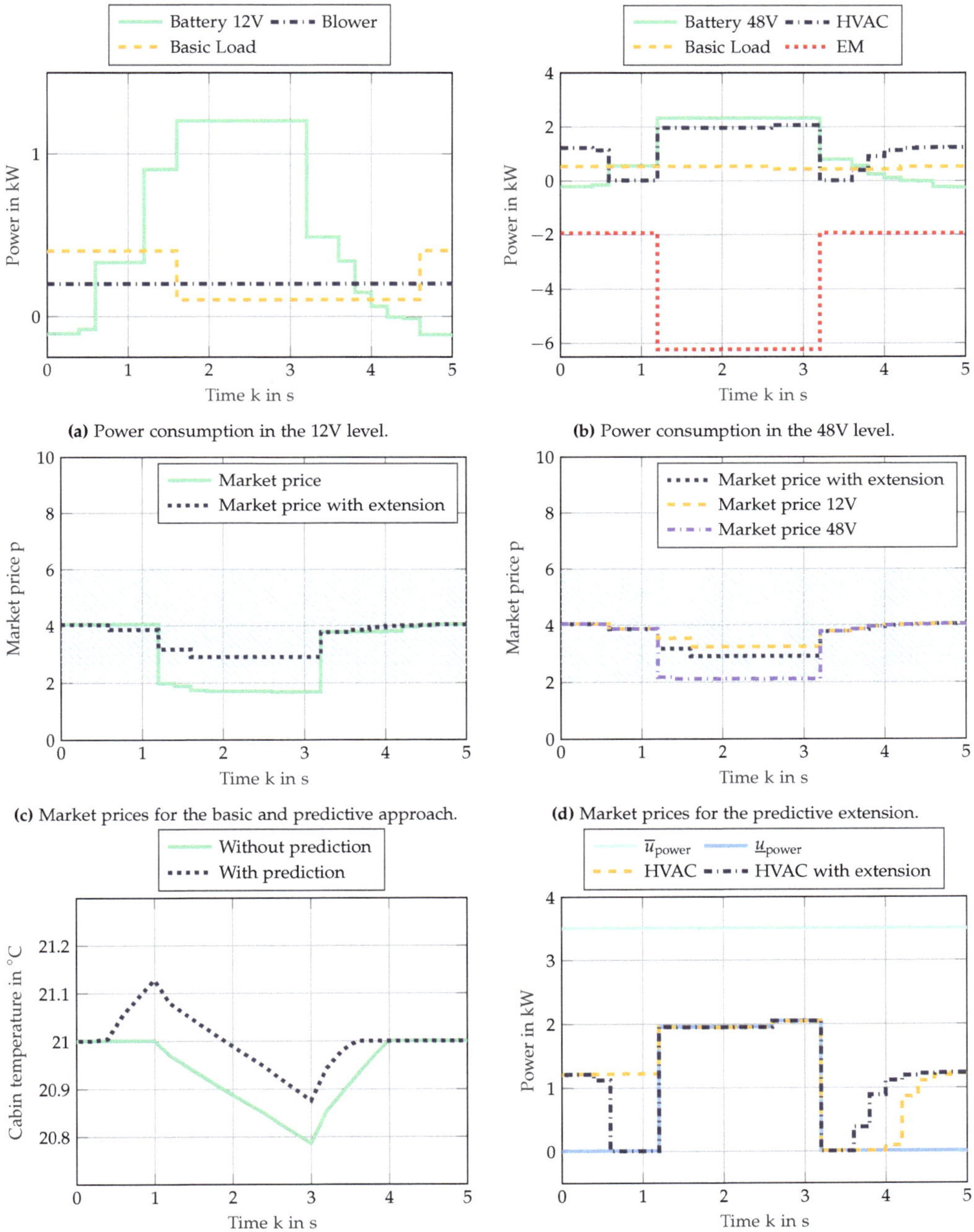

(a) Power consumption in the 12V level.

(b) Power consumption in the 48V level.

(c) Market prices for the basic and predictive approach.

(d) Market prices for the predictive extension.

(e) Cabin temperature.

(f) Power consumption of the HVAC system.

Figure 5.20: Simulation results from the second scenario with the EM in recuperation mode, the activated predictive extension, and limitations in terms of PE power transfer. The predictive extension keeps the market price within the normal price interval Δp_{norm} marked by the blue stripes.

level is cut at $\hat{P}_{\text{PE, lim}}^{12\text{-to-}48} = 1000\,\text{W}$, and the transfer is cut at $\hat{P}_{\text{PE, lim}}^{48\text{-to-}12} = 1500\,\text{W}$ for the reverse direction as listed in Table 5.2. Figure 5.19 and Figure 5.20 show the simulation results for the two scenarios, including the active predictive extension. The individual graphs strongly distinguish from the results of the basic approach in Figure 5.7 and Figure 5.8. The differences between the two approaches are also visible in the market prices. In Figure 5.19c, the deviating prices are visualized for the first scenario with the EM in boost mode. While the market price for the basic approach in green leaves the normal price range $\Delta p_{\text{norm}} = (2,6]$, the predictive extension achieves a market price within this price range.

During both scenarios, the market price for the approach with the predictive extension given by the blue dashed graph stays in the normal price interval $\Delta p_{\text{norm}} = (2,6]$. Due to the load shifting, the market prices again differ directly before and after the boosting and recuperation phase in comparison to the basic approach given by the green market price graph. In these periods, the HVAC system compensates for the exaggerated or reduced consumption during the phases of power shortening or power surplus. The load shifting, which is controlled by the MPC of the HVAC system, is pointed out in the blue dashed graph in Figure 5.19f and in Figure 5.20f. As a consequence of the adapted power consumption scheme, the cabin temperature T_{cab} in Figure 5.19e and in Figure 5.20e deviates from the desired temperature $T_{\text{cab, ref}} = 21\,°\text{C}$. Nevertheless, if compared to the temperature deviation for the basic approach without the predictive extension in green, the highest absolute difference to the reference temperature $T_{\text{cab, ref}}$ in Figure 5.20e is reduced for the activated predictive extension.

The PE limitations are another aspect of the simulations. As depicted in Figure 5.19d and in Figure 5.20d, the market prices in the power levels differ due to the lower PE limitations. The two voltage levels are considered one market without the PE limitations since the DC/DC-converter transfers power accordingly. But the PE limitations put constraints on the power transfer, which can not be solved by the predictive extension and lead to separated power markets. As a result, during the extremes in the boosting and recuperation period, the power transfer is limited, leading to deviating price levels for the 12V and the 48V levels. Nevertheless, in comparison with the basic approach in Figure 5.7 and Figure 5.8, the global market price and the individual market prices for the voltage levels are mostly kept within the normal range.

An exception is shown in Figure 5.19d during the boosting phase where the market price 48V in yellow exceeds the normal market price range $\Delta p_{\text{norm}} = (2,6]$ marked with the blue stripes. The detailed view of the different market prices points out the limitations of the predictive extension. In Figure 5.19c and in Figure 5.19d, the global market price for predictive extension in blue stays inside $\Delta p_{\text{norm}} = (2,6]$. However, in Figure 5.19d, at the beginning of the boosting phase from about $k = 1.5\,\text{s}$ to $k = 2.5\,\text{s}$, the market price for the 48V level displayed by the yellow dashed graph exceeds the normal price range $\Delta p_{\text{norm}} = (2,6]$. Consequently, the EM is forced to constrain the boosting power to $P = 4500\,\text{W}$, highlighted in the red dotted graph in Figure 5.19b. The power support by the 12V level is limited through the PE limitations $\hat{P}_{\text{PE, lim}}^{12\text{-to-}48}$, leading to a lower market price for the 12V level marked by the purple dashed graph in Figure 5.19d and a higher market price for the 48V level in yellow. Figure 5.19f demonstrates the load shifting of the HVAC system to meet the predicted power constraints. However, during the boosting phase from about $k = 1.5\,\text{s}$ to $k = 2.5\,\text{s}$, in which the power deficit occurs, the HVAC system is already reducing its power consumption to zero. Thus, the HVAC system has no further potential for additional load shifting in this period.

During the recuperation phase in the second scenario, the power transfer limitation toward the 12V level plays a less significant role. The MPC in the HVAC system compensates for the additional electric power that can not be transferred to the 12V level by shifting more power consumption in the period of regenerative braking as depicted in Figure 5.20f. Thus, the HVAC system increases the power consumption during the regenerative braking, directly consuming the available power. In contrast, there is a phase of reduced consumption before the recuperation period shown by the blue dashed graph and compared to the auction mechanism, which is given by the yellow dashed graph in Figure 5.20f. Consequently, both batteries are charging before and after the recuperation phase in Figure 5.20 since the HVAC system decreases the cooling to zero in these intervals.

5.6.2 Capabilities and Limitations

To have a detailed view of the capabilities and limitations of the predictive extension, the simulation with a reduced $SOC_{48V} = 0.4$ for the Battery 48V is performed with the extended approach. Again, the simulation corresponds to the previous description regarding the first scenario and the parameters in Table 5.1. The only differing aspect is the new SOC, which results in the battery 48V PPF representation in Figure 5.12 marked by the green dashed graph.

Since the whole PPF shifts to the right side toward higher market prices, the battery 48V has a diminished power supply behavior in the normal price interval $\Delta p_{\text{norm}} = (2, 6]$. As visualized in Figure 5.12, the battery 48V will charge until a market price $p > 6$ is reached. Thus, the battery is mainly charges to restore the SOC and only contributes to the overall power supply if the power network needs additional power.

Compared to the simulation results in Figure 5.13, only in the phase of boosting activity is the battery discharging to support the power supply and enable the EM boosting. During the rest of the simulation in Figure 5.21, the battery in the 48V level is mainly charging, resulting in a higher demand for the battery 12V which supports the 48V level via the DC/DC-converter as depicted in Figure 5.21a. Additionally, the HVAC system in Figure 5.21d performs a more intense load shifting if compared to the previous simulation with the auction-based power management approach in Figure 5.13d.

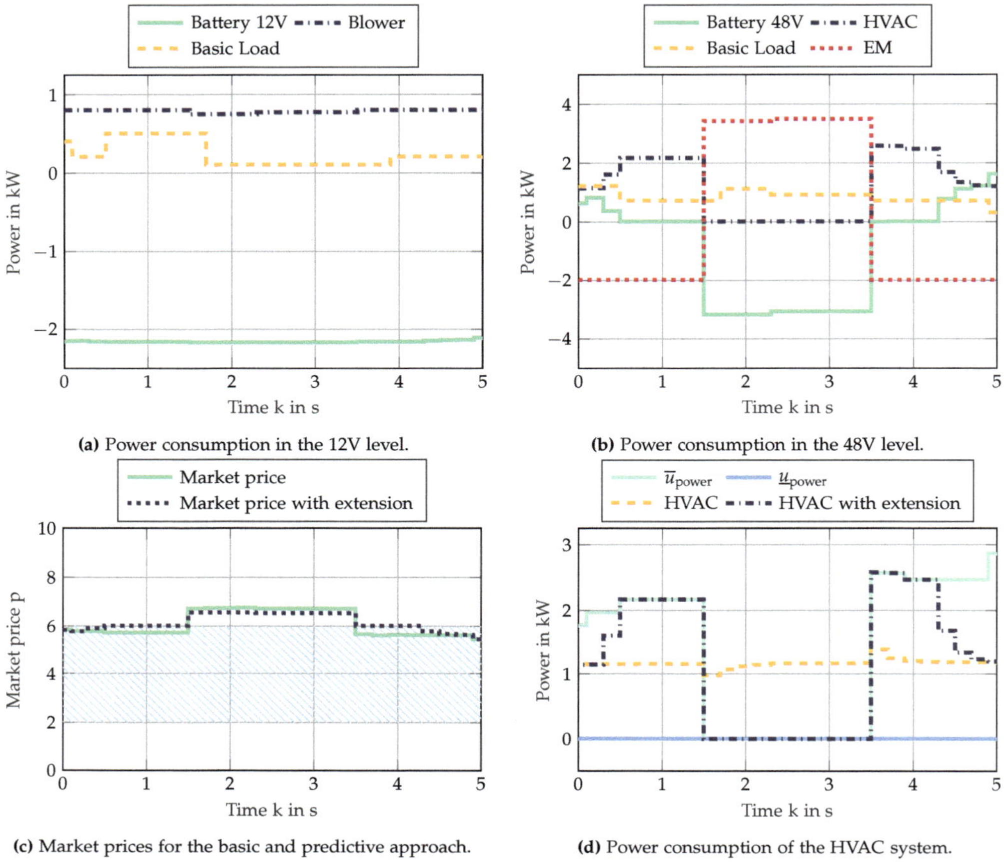

(a) Power consumption in the 12V level.

(b) Power consumption in the 48V level.

(c) Market prices for the basic and predictive approach.

(d) Power consumption of the HVAC system.

Figure 5.21: Simulation results from the first scenario with the EM in boost mode, the activated predictive extension, and a battery 48V $SOC_{48V} = 0.4$. The results demonstrate the limitations of the predictive extension.

The load shifting differs from the previous results, in particular, due to the upper power constraint $\overline{u}_{power}$ shown in the light green graph. Since the battery 48V with an $SOC_{48V} = 0.4$ has a great influence on the power level at a market price of $p = 6$, the upper power constraint $\overline{u}_{power}$ for the MPC is mostly lower than the maximum power consumption of the HVAC $P_{max,HVAC} = 3500\,W$ (see Table 5.1). As a consequence, the HVAC's load shifting has to be spread over a more extended period.

Because of the limited power support of the battery 48V, the load shifting during the boosting phase is insufficient to allow for the desired boosting power $P_{set} = 4750$ W. Instead, the EM is forced to reduce the boosting power to about $P = 3750\,W$ as depicted in Figure 5.21b by the red dotted graph. The effect is also visible in the market price for the predictive extension displayed by the blue dotted graph in Figure 5.21c. Despite the exploitation of the HVAC system, the market price exceeds the normal price interval $\Delta p_{\text{norm}} = (2, 6]$ during the whole boosting phase. In Figure 5.21a, the effect on the blower in the blue dashed graph is visualized. Due to the full exploitation of the battery 12V and the increased market price, the blower is forced to slightly limit its power consumption to balance the overall power in the APN.

In the simulation with the added basic loads in the 12V and the 48V level, the HVAC system is again fully exploited as visualized in Figure 5.22d. Still, the boosting activity has to be reduced from $P_{\text{set}} = 4750\,\text{W}$ to about $P = 4500\,\text{W}$. The power consumption and supply during the scenario for the 12V and for the 48V level are displayed in Figure 5.22a and in Figure 5.22b, respectively. The new market price in Figure 5.22c slightly exceeds the normal price interval compared to the market price regarding the basic approach in green.

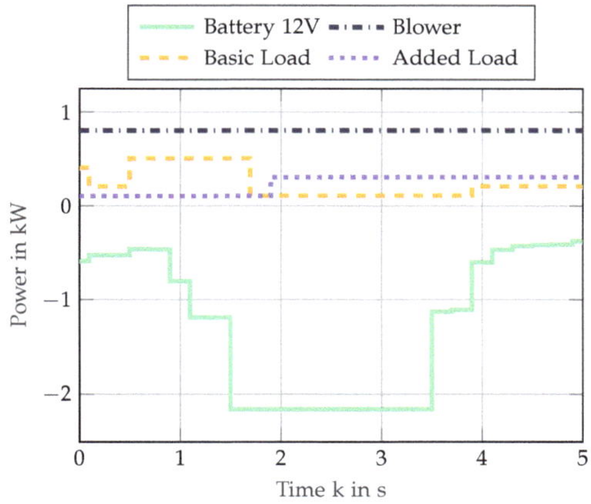

(a) Power consumption in the 12V level.

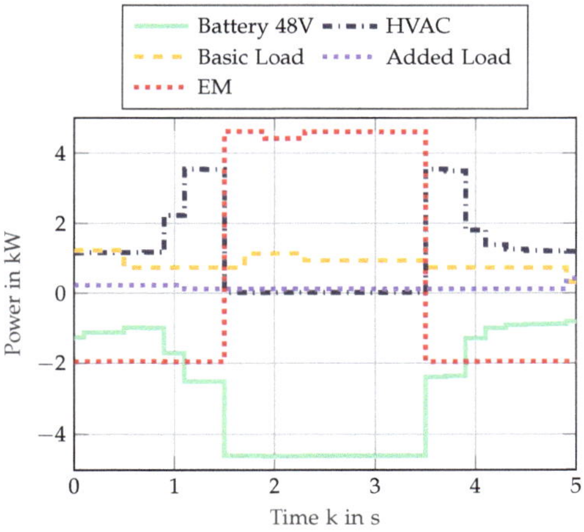

(b) Power consumption in the 48V level.

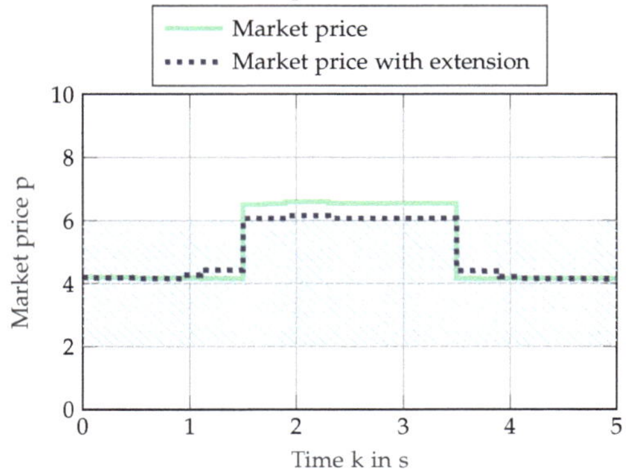

(c) Market prices for the basic and predictive approach.

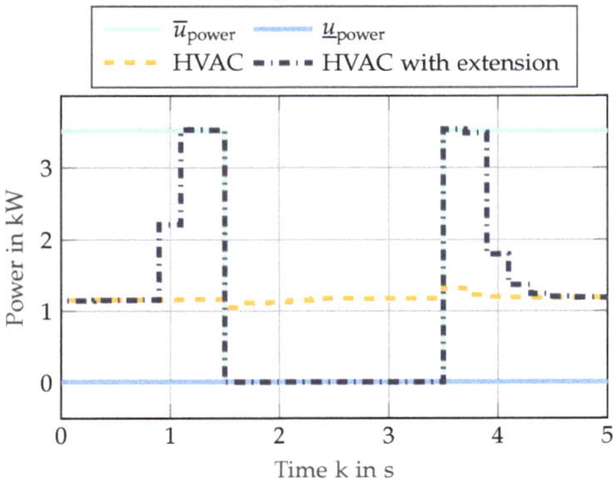

(d) Power consumption of the HVAC system.

Figure 5.22: Simulation results from the first scenario with the EM in boost mode, the activated predictive extension, and added basic loads in both power levels.

5.6.3 Discussion of the Predictive Extension

The predictive extension adds advantages to the auction-based power management. In particular, the predictive extension keeps the market price mainly within the normal price range Δp_{norm}. Hence, the predictive load shifting increases the power balance capability, preventing other comfort loads from shutting down or mitigating battery strain. The exemplary implementation for the HVAC system brings a high potential for load shifting through the different heat storages and considerable time constants [KLFE11, VA15].

The applied MPC algorithm is a capable methodology to consider the power constraints $\overline{u}_{\text{power}}$ and $\underline{u}_{\text{power}}$ and, at the same time, enhance HVAC system performance. Since the current implementation emphasizes the basic functionality, there are still MPC features that are not yet included but may offer further improvements. For example, the power gradients may be adapted to allow smooth transients and mitigate wearing effects through abrupt power changes. Further, the MPC algorithm supports the consideration of individual passenger preferences with regard to the acceptable temperature deviations [LLDL11]. Hence, the load-shifting potential grows by increasing the allowable temperature difference $|\Delta T_{\text{cab, ref}}|$. Moreover, the MPC allows for considering maximal deviations $|\Delta T_{\text{cab, ref}}|$ from the desired cabin temperature.

A restriction of the predictive extension is the dependence on flexible comfort loads. Furthermore, these loads are only helpful if they are active and, thus, can be controlled in their current power demand to shift future loads. Consequently, this work focuses on the HVAC system, which is usually active to control the temperature and the air condition in the vehicle interior. Additional flexible comfort loads may be added to a pool of possible components for load shifting. However, the current implementation only regards one component and does not facilitate the utilization and coordination of multiple loads to increase the load-shifting potential. However, the availability of flexible comfort loads for most vehicles is limited to the HVAC system and the seat heating.

For the effectiveness of the MPC algorithm, a precise and reliable prediction is essential [BBM17, HWK+17, RSSH21]. Therefore, the APN components have to provide the necessary information for the regarded prediction horizon N_p to predict their power consumption and, subsequently, calculate adequate control inputs. Since the auction-based approach focuses on the comfort loads, the main uncertainty, in this context, is the passenger behavior. Even though passenger behavior is not deterministic and highly individual, there are approaches and investigations to forecast possible passenger actions [GLS+20].

Another factor, referring to the comfort loads, especially the HVAC system, is the weather and the ambient temperature T_{amb}. As described in Section 4.2, the HVAC system behavior directly depends on the ambient conditions [KLFE11]. On the one hand, the weather and ambient temperature influence the in-vehicle temperature. On the other hand, the outdoor conditions influence the effectiveness of the HVAC system in terms of cooling, heating, or air conditioning. In this regard, even the humidity plays a significant role with respect to the power consumption [KLFE11, Ste14].

Weather conditions also influence driving operation and safety. The road surface characteristics influencing the tires' friction factor strongly impact driving and stability. Further, the driving operation depends on human decisions and driving behavior, which is only predictable with uncertainty. Accordingly, the growing dissemination of assisting and autonomous driving functions offers a reliable driving load prediction [ORBE20, RSSH21]. A currently available source may be the navigation system which includes road profiles and slope information.

As long as these uncertainties remain, the proposed predictive extension needs to cope with the prevailing information and the uncertainties. Since the MPC recalculates the input trajectory for every new instant of time, there is a recurrent opportunity to correct and improve the last prediction [BBM17]. Consequently, the predictive extension is already capable of partly overcoming this limitation.

5.7 Summarizing Discussion

This section gives an overview of the key results from Chapter 4 and Chapter 5. The accomplishments and results are recapitulated and discussed based on the defined properties for a capable automotive power management approach in Section 2.2.

The holistic power balancing by the auction-based automotive power management and, in particular, the proposed predictive extension ensure a high *quality of power balancing*. The simulation results demonstrate the working principle and the effectiveness of the proposed power management in challenging situations with strongly volatile power supply and demand. Additionally, the predictive extension adds further power-balancing capabilities by exploiting flexible comfort loads.

Regarding the desired *plug-and-play integration* of new hardware components into the APN, the auction-based mechanism in combination with the SOA based communication design is a favorable solution. Despite the auctioneer, which represents a centralized part of the auction mechanism, the overall procedure facilitates the seamless integration of new components into the groups of sellers and bidders. Since the auctioneer is a centralized service in the SOA, which is reachable by all components, it represents a mutual marketplace for the power auction in the APN. On the other hand, the current implementation of the predictive extension is designed for one centralized element, which is the HVAC system as a flexible comfort load. However, this centralized extension does not impair the general plug-and-play integration of new hardware components.

In terms of *scalability*, the proposed power management mechanism provides noteworthy advantages. New components are easily added to the cumulative PPF. Subsequently, the components represented by their PPF in the PPF_{cum} are considered in the power balancing process within the UPA. Hence, the computational effort for every additional participant in the auction mechanism is neglectable.

One major drawback of the basic idea in [Gra04a] is the poor *generalization* for APNs with multiple voltage levels. By the extension for multiple voltage levels and introducing PPFs with mandatory design rules, the contributions in this work improve the auction-based power management toward the application in future APNs.

The *individualization and customization* for the auction-based mechanism is facilitated by the flexible design of the PPFs within the design rules given in Table 4.1. On the one hand, the PPFs comprise component states and information that influence the individual behavior in the UPA as highlighted for the SOC_{48V} in Section 5.4. On the other hand, user preferences may be included in the PPFs. Consequently, the PPFs support distributed decision-making and design while the UPA Definition 4.1 and the design rules in Table 4.1 guarantee a robust and reliable power balancing framework.

Due to the need for the auctioneer, the centralized auction-based design is inherently less fail-safe than comparable distributed approaches [Ise16, CSAH23]. Nevertheless, through the combination with the SOA, the proposed mechanism offers *fail-safety* for a single point of failure [SGLS22, HSS+22]. In case of failure, a second auctioneer instance that shadows the primary auctioneer service overtakes the functionality, as discussed in the communication concept in Section 4.3. Thus, the flexible instantiation of services among the different computational units in the SOA is the key factor for enhanced fail-safety [RGKS20, SGLS22]. Further, the SOA communication patterns facilitate efficient communication and the plug-and-play integration of new hardware components.

Regarding the *computational effort*, the proposed auction-based algorithm causes a neglectable load, comparable to rule-based approaches [KET+11, RBW+12]. In contrast, the implementation of the predictive extension causes additional computational effort for the MPC and the prediction of overall power consumption in the APN. However, the availability of computational power in cars is steadily growing, which may provide the necessary resources for the proposed predictive extension [ORBE20].

Within the auction process and for the individual PPF design, there is a certain *communicational effort* that has to be considered in the APN design in terms of communication technology. The auction-based power management utilizes the information already broadcasted in the vehicle, such as the component states, reducing additional communication and computation effort. Since the communication based on the SOA is a critical element of the proposed power management, the communication design has to be carefully considered for implementation.

In terms of simplicity and *costs of implementation*, the auction-based mechanism for power management in the APN has superior advantages. The auction design provides a clear structure, and the possible customization is shifted toward the individual PPFs of the components facilitating a distributed design. In turn, the individual PPF design follows the guidelines of the fixed and mandatory market price ranges and the rules within the UPA definition. Hence, the auction-based approach offers a clear and explicit structure toward a capable, easy-to-implement power management framework.

For the predictive extension, various algorithms and software components have to be implemented. These components comprise a precise HVAC model and the MPC algorithm, which has to be tuned for high control performance in the specific vehicle. As a result, the predictive extension requires expertise and supplemental implementation effort. However, these implementation endeavors pay off through the improved quality of power balancing.

In conclusion, the formalized auction-based power management with its predictive extension achieves reliable power balancing for vehicles. Simultaneously, it enables the desired plug-and-play integration of new hardware components. By the UPA definition and the design rules for PPFs, the approach is mathematically formalized for the first time. Due to the generalization for multiple voltage levels, the auction-based power management approach is suitable for the application in APNs. Furthermore, with the predictive extension, the quality of power balancing and, thereby, the mitigation of battery strain is further enhanced.

6 Conclusion

Due to the ongoing electrification, the dissemination of entertainment and infotainment systems, and the broader application of assisting and autonomous driving functions, the APN faces a growing number of electric and electronic components and, thereby, rising power demands. The automotive industry needs a shift in hardware, software, and communication concepts to meet these challenges, ensuring a stable power supply and the flexibility for plug-and-play integration of new hardware components during development and even after production. In this work, an automotive power management approach tackling these desirable properties is investigated and developed.

First, state-of-the-art concepts for automotive power management, as well as the methodologies and ideas in neighboring domains, are investigated. The approaches are divided into four categories: rule-based, dynamic optimization-based, agent-based, and auction-based. The qualitative assessment focuses on power balancing quality, flexibility concerning plug-and-play integration, and further properties of capable power management. The concluding comparison provides a suitable starting point for fruitful investigations and encourages the auction mechanism as a promising idea.

Therefore, this work formalizes and extends an auction-based power management approach that suits the voltage level architectures, the related hardware components, and the communication technologies of modern vehicles. The approach includes PPFs to describe the individual component behavior and operation within the auction. Furthermore, the auction procedure considers multiple voltage levels in the MCP determination and a hysteresis mechanism for stepwise switching loads. The proposed power management is complemented with a predictive extension that exploits flexible comfort loads for intelligent load shifting to enhance the power balancing capability. The combination with a communication design based on a SOA yields the plug-and-play integration property and enhances fail-safe operation.

The proposed power management is evaluated in simulations of relevant driving scenarios. The simulation results highlight the effectiveness of the auction-based working principle for power balancing in the APN during challenging situations with strongly volatile power demands. Furthermore, the results demonstrate the generalization for multiple voltage levels, the effect of stepwise switching loads, and the seamless integration of new loads into the auction procedure. The power balancing capabilities are further enhanced by applying the predictive extension since it prevents unintentional load reduction. In the summarizing discussion, the proposed auction-based power management is comprehensively assessed regarding the defined properties of capable automotive power management.

While this work has shown the general working principle with few components in two voltage levels, extended simulations evaluating the integration of more electric and electronic components are required. An essential step toward real application is the integration of the auction mechanism into the automotive software framework. In particular, the communication process and the synchronization between all auction participants are critical for a successful implementation. Afterward, the transfer toward test benches or conceptual vehicles is the next step in order to validate the interconnection with other systems and the robustness against communication issues, such as delays or loss of information. One crucial aspect is the interface design to the electric and electronic components in the vehicle. Since the development and production of most components are outsourced to suppliers in the automotive industry, the distributed auction mechanism is challenging to deploy, and comprehensive coordination of software interfaces is required. Furthermore, the approach needs robustness in terms of measurement errors of the components and errors in the load prediction. However, the passive balancing capabilities of the batteries and other components in the power network compensate for the active power balancing impreciseness, including the PPF hysteresis design.

Regarding the salient generalization toward multiple voltage levels and the scalability in terms of participants, auction-based power management may be a viable solution for other mobility systems, such as buses, trucks, or ships. For these systems, the plug-and-play integration of electric and electronic hardware components provides even more advantages than for the automobile. Hence, the proposed auction-based power management concept is promising in different transportation systems.

Another possible field of application for the proposed auction-based mechanism is energy management. However, compared to auction mechanisms, centralized optimization approaches account explicitly for multiple objectives, such as efficiency in driving operations and in the electric domain. Therefore, centralized optimization approaches may be more suitable and effective than the auction-based procedure concerning the automotive energy management shell. Since the auction-based procedure emphasizes balancing and distributing resources, it neglects the consideration of specific optimization goals.

Through the promising simulation results in this work and the outlined steps in the validation process, the methodology is ready for application in future vehicles and may contribute to a flexible and sustainable vehicle platform and a reduction in development efforts.

A Detailed Heating, Ventilation, and Air Conditioning Model

The heating, ventilation, and air conditioning (HVAC) system description in Section 4.2 provides an overview of the model and the underlying thermodynamic phenomenons. Nevertheless, the comprehensive HVAC model includes various components and their heat exchange. Since these relations are difficult to present concisely, the exact and profound modeling is shown in this chapter. Furthermore, Section A.3 lists the relevant parameters for the HVAC model concerning the presented simulation results in Section 5.6. The following sections are based on the extended information in [KLFE11, Ste14].

A.1 Equivalent Circuit Diagram and Thermodynamic Mechanisms

In Figure A.1, the thermal relations between the different components in direct proximity to the passenger compartment are presented within an equivalent circuit diagram. The depicted elements comprise the cabin air, the passengers as heat sources, the interior equipment and components, the vehicle roof, and the windows. In parts, the components are irradiated by the sun which is depicted with a current source and the heat input $\dot{Q}_{\text{solar}}$. On the other hand, the voltage sources describe the heat convection to the ambient T_{amb} or the heat input by the passengers T_{pers}. The heat flow results from the temperature difference and the respective resistors. The current source on the left side highlights the heat input from the HVAC system $\dot{Q}_{\text{HVAC}}$.

Nodes in the diagram mark temperature levels, listed in the state vector $x(t)$, and partly given in the figure. The different temperature levels in the vehicle are separated by resistors R representing the heat convection between the respective components. In this regard, the resistors with the index i represent the convection toward the cabin air, and the index a represents the convection to the ambient. The connecting temperature on the top line is the cabin air temperature T_{cab}. Conversely, the bottom line marks the temperature level $T = 0\,°\text{C}$. The heat capacity of the cabin air C_{air} stabilizes this temperature level T_{cab} by storing heat emitted if the surroundings are colder than the cabin air.

The thermal resistors R_{conv} describing the heat convection are calculated by

$$R_{\text{conv}} = \frac{1}{\alpha A_{\text{conv}}} \tag{A.1}$$

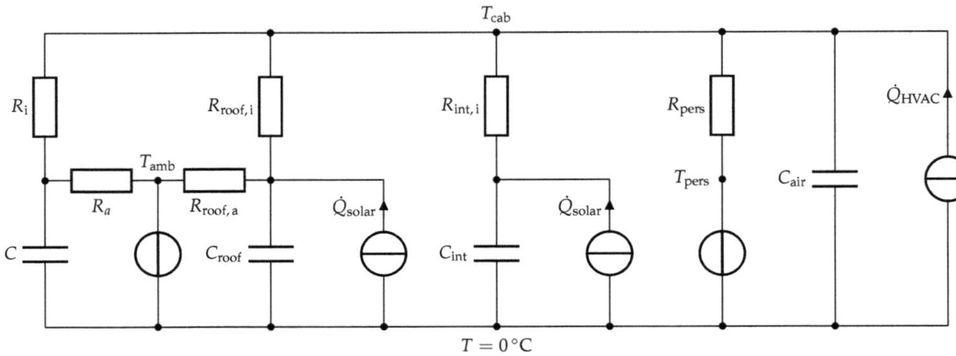

Figure A.1: Equivalent circuit diagram for the modeled HVAC system with its various components and the considered heat exchange.

and include the coefficient for heat transfer α and the component surface A_{conv} which is involved in the heat convection. The coefficient for the inner convection is given by

$$\alpha_{\text{in}} = \max\left\{7\frac{\text{W}}{\text{m}^2\text{K}}; \kappa_{\text{in}}\sqrt{\dot{V}_{\text{air}}}\right\}. \tag{A.2}$$

The volume flow $\dot{V}_{\text{air}}$ has the unit m^3/h and represents the airflow that is controlled by the HVAC system. On the other hand, the factor κ_{in} describes the component-specific heat convection. The minimal value for the coefficient regarding inner convection is set to $7\,\frac{\text{W}}{\text{m}^2\text{K}}$ [KLFE11]. The convection coefficients for the respective components are listed in Table A.1. To consider the thermal insulation of the roof and the body elements, the coefficient κ_{in} for these components is set to a fixed value of $2.3\,\text{W}/\text{m}^2\text{K}$ [KLFE11, p. 21].

The heat convection to the ambient depends on the vehicle velocity v_{car} in m/s and is approximately described by

$$\alpha_{\text{amb}} = \max\left\{25\,\frac{\text{W}}{\text{m}^2\text{K}}; \kappa_{\text{amb}}\, v_{\text{car}}^{0.8}\right\}. \tag{A.3}$$

The coefficients κ_{amb} for the relevant components are listed in Table A.1. With regard to slow velocities, the minimal value for α_{amb} is set to $25\,\frac{\text{W}}{\text{m}^2\text{K}}$ [KLFE11, p. 74].

The thermal resistor of a person is described by

$$R_{\text{pers}} = \frac{1}{\alpha_{\text{pers}} A_{\text{pers}}}. \tag{A.4}$$

Accordingly, the heat input of the passengers is modeled as a heat source with a fixed body temperature of $T_{\text{pers}} = 36\,°\text{C}$ and the resistor R_{pers}. With reference to [BBGV15], the coefficient α_{pers} is set to $5\,\text{W}/\text{K}$. With a heat input under normal circumstances of $70\,\text{W}/\text{m}^2$ and an approximate body surface of $A_{\text{pers}} = 1.8\,\text{m}^2$ for an adult, the heat input considered in the HVAC model is about 125 W per person.

Table A.1: Parameters κ for the heat convection to the ambient (amb) and to the interior (in).

Component	κ_{amb}	κ_{in}
Roof	4.41	–
Body	7.21	–
Front windshield	3.79	0.584
Rear windshield	4.65	0.7
Side windows	7.21	0.495

A.2 Underlying Control Loop and State Space Model

The HVAC system takes the warm ambient air and cools it down in order to cool the cabin air and provide fresh air for the passengers. Since the warm air from outside contains more water steam, the HVAC system causes condensation of water at the evaporator. Thus, the HVAC has to provide additional cooling for the condensation. This latent heat $\dot{H}_{\text{latent}}$ is expressed by

$$\dot{H}_{\text{latent}} = r\dot{m}_{\text{cond}} \tag{A.5}$$

and includes the evaporation heat $r = 2500\,\text{kJ/kg}$ and the mass flow $\dot{m}_{\text{cond}}$. The condensate mass flow

$$\dot{m}_{\text{cond}} = \dot{m}_{\text{air}}(x_{\text{amb}} - x_{\text{HVAC}}) \tag{A.6}$$

depends on the mass flow of the air from outside $\dot{m}_a ir$ and the difference between the relative humidity outside x_{amb} and after the cooling process x_{HVAC}.

The sensible heat that is necessary to cool down the air from the ambient temperature T_{amb} to the temperature after the evaporator is calculated by

$$\dot{H}_{\text{sens}} = \dot{m}_{\text{air}} c_{\text{air}} (T_{\text{amb}} - T_{\text{HVAC}}). \tag{A.7}$$

According to [GB20], the enthalpy of the evaporator, which includes the latent heat and the sensible heat, is given by

$$\dot{H}_{\text{evap}} = \dot{m}_{\text{air}}(h_{\text{amb}} - h_{\text{HVAC}}). \tag{A.8}$$

The specific enthalpy of the air after the evaporator h_{HVAC} is set to $18.6\,\text{kJ/kg}$ [Ste14]. Regarding the air after the evaporator, the temperature is assumed to be $T = 5\,°\text{C}$, and the relative humidity is set to 100 %. As described in [GB20], the temperature at the evaporator is kept between $T = 5\,°\text{C}$ and $T = 10\,°\text{C}$ in order to prevent ice formation. Hence, the HVAC cooling activity is controlled by the airflow $\dot{m}_{\text{air}}$ toward the cabin. The necessary air flow depends on the set cooling input $\dot{Q}_{\text{HVAC}}$ and holds

$$\dot{m}_{\text{air}} = \frac{\dot{Q}_{\text{HVAC}}}{c_{\text{air}}(T_{\text{cab}} - 5\,°\text{C})}. \tag{A.9}$$

A PI-controller is utilized to control and stabilize the cabin temperature T_{cab}. The PI-controller calculates the control error

$$e(t) = T_{\text{ref}}(t) - T_{\text{cab}}(t) \tag{A.10}$$

and derives the necessary heat transfer $\dot{Q}_{\text{HVAC}}$ to cool the cabin temperature and minimize the control error $e(t)$. With reference to the following state space model, the PI-controller adds a state for the integral part, which is given by

$$\dot{x}_i = K_i(T_{\text{ref}} - T_{\text{cab}}). \tag{A.11}$$

Here, K_i is the gain of the PI-controller for the integral part. With the proportional part and its gain K_p, the cooling input

$$\dot{Q}_{\text{HVAC}} = x_i + K_p(T_{\text{ref}} - T_{\text{cab}}) \tag{A.12}$$

is calculated. Consequently, the cooling input is derived from the states x_i and T_{cab} and from the reference temperature T_{ref}, which is set by the MPC algorithm. To describe the controller delay, a PT1-element is integrated, which slows down the control input with the time constant $T_p = 0.04\,\text{s}$.

In summary, the HVAC system behavior is expressed by the state space model

$$\dot{x}(t) = \mathbf{A}x(t) + \mathbf{B}u(t) \tag{A.13}$$

$$y(t) = \mathbf{C}x(t) + \mathbf{D}u(t). \tag{A.14}$$

The states $x(t)$ in (4.27) include the temperatures in the equivalent circuit diagram in Figure A.1. These temperatures are given at the different nodes in the equivalent circuit diagram and represent the temperatures of the components in the vehicle. Additionally, the cabin air temperature T_{cab} is a state that is connected to all components through heat convection. These connections are depicted in the matrix

$$\mathbf{A} = \begin{bmatrix} -\frac{1}{C_{\text{front}}}\left(\frac{1}{R_{\text{front,i}}} + \frac{1}{R_{\text{front,a}}}\right) & 0 & \cdots & \frac{1}{R_{\text{front,i}}C_{\text{front}}} \\ 0 & \ddots & 0 & \vdots \\ \vdots & 0 & \ddots & \vdots \\ \frac{1}{R_{\text{front,i}}C_{\text{cab}}} & \frac{1}{R_{\text{side,i}}C_{\text{cab}}} & \cdots & -\frac{1}{C_{\text{cab}}}\left(\frac{1}{R_{\text{front,i}}} + \cdots + \frac{1}{R_{\text{pers}}}\right) \end{bmatrix}. \tag{A.15}$$

On the other hand, the inputs $u(t)$ in (4.28) comprise the ambient temperature T_{amb}, the solar radiation E_{solar}, the temperature of the passengers T_{pers}, and the input from the HVAC system $\dot{Q}_{\text{HVAC}}$. The inputs' influence on the states $x(t)$ is described by the matrix

$$\mathbf{B} = \begin{bmatrix} \frac{1}{R_{\text{front, a}} C_{\text{front}}} & 0 & 0 & 0 \\ \frac{1}{R_{\text{side, a}} C_{side}} & 0 & 0 & 0 \\ \frac{1}{R_{\text{rear, a}} C_{\text{rear}}} & 0 & 0 & 0 \\ \frac{1}{R_{\text{body, a}} C_{\text{body}}} & 0 & 0 & 0 \\ \frac{1}{R_{\text{roof, a}} C_{\text{roof}}} & \frac{A_{\text{roof}}}{C_{\text{roof}}} & 0 & 0 \\ 0 & \frac{0.15\, A_{\text{int}}}{C_{\text{int}}} & 0 & 0 \\ 0 & 0 & \frac{1}{R_{\text{pers}} C_{\text{cab}}} & -\frac{1}{C_{\text{cab}}} \end{bmatrix} . \tag{A.16}$$

With the output matrix

$$\mathbf{C}^T = \begin{pmatrix} 0 & 0 & 0 & 0 & 0 & 0 & 1 \end{pmatrix} . \tag{A.17}$$

the cabin temperature T_{cab} is defined as the system output. The prevailing system's feedthrough matrix $\mathbf{D}$ is zero.

To translate the necessary heat input $\dot{Q}_{\text{HVAC}}$ into an electric power P_{comp} of the compressor, the

$$\text{COP} = \frac{\dot{Q}_{\text{HVAC}}}{P_{\text{comp}}} \tag{A.18}$$

approximates the degree of efficiency [Ste14]. In this regard, the coefficient of performance (COP), which depends on the ambient temperature T_{amb}, connects the heat flow for the HVAC cooling and the corresponding electric power running the cooling circuit. For the COP in Figure A.2, the evaporator temperature is set to a constant value of $T = 5\,°\text{C}$. This assumption has already been used in (A.9) to calculate the necessary airflow $\dot{m}_{\text{air}}$.

A.3 Parameters

Regarding the simulations shown in Section 5.6, the following parameters for the components in the HVAC model are used. The values represent a middle-class vehicle and are provided in [KLFE11]. Table A.2 lists the parameters describing the thermal behavior for the included components.

Additionally, Table A.3 shows further parameters concerning the HVAC system and the surrounding components. These parameters are necessary to set up the simulation environment appropriately.

Table A.2: Parameters of the vehicle components for the thermodynamic model and the calculation of the heat capacities [KLFE11].

Component	Variable	Value
Passenger compartment	Volume	$2.5\,\mathrm{m}^3$
	Mass	13 kg
Roof	Specific heat capacity	480 J/(kg K)
	Surface	$1.7\,\mathrm{m}^2$
	Mass	41 kg
Side wall	Specific heat capacity	480 J/(kg K)
	Surface	$3.4\,\mathrm{m}^2$
	Mass	12 kg
Splash back wall	Specific heat capacity	480 J/(kg K)
	Surface	$1.1\,\mathrm{m}^2$
	Mass	33 kg
Underside	Specific heat capacity	480 J/(kg K)
	Surface	$2.8\,\mathrm{m}^2$
	Mass	12.1 kg
Front windshield	Specific heat capacity	804 J/(kg K)
	Surface	$1.1\,\mathrm{m}^2$
	Mass	4.8 kg
Rear windshield	Specific heat capacity	804 J/(kg K)
	Surface	$0.5\,\mathrm{m}^2$
	Mass	11.4 kg
Side windows	Specific heat capacity	480 J/(kg K)
	Surface	$1.2\,\mathrm{m}^2$
	Mass	300 kg
Interior equipment	Specific heat capacity	1250 J/(kg K)
	Surface	$9\,\mathrm{m}^2$

Table A.3: Further parameters for the HVAC model and the surrounding systems.

Description	Variable	Value
Integral gain	K_i	-10
Proportional gain	K_p	-200
Time constant of the PT1-element	T_p	0.04 s
Percentage of irradiated components	–	15 %
Temperature of a passenger	T_{pers}	36 °C
Number of passengers	n_{pers}	1
Assumed evaporator temperature	T_{evap}	5 °C
Passenger heat input	–	125 W

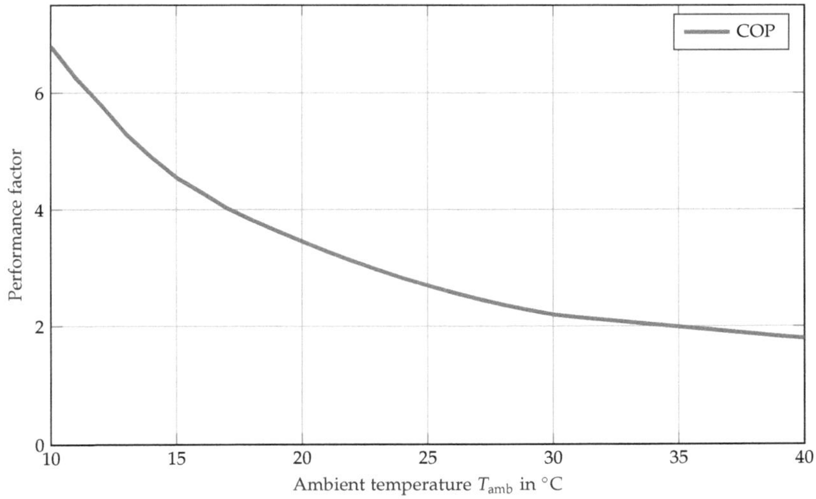

Figure A.2: COP of the HVAC system depicting the relation between the HVAC efficiency and the ambient temperature T_{amb} [Ste14].

References

Public References

[ABJG12] AYEB, M.; BRABETZ, L.; JILWAN, G.; GRAEBEL, P.: A Generic Modeling Approach for Automotive Power Net Consumers. In: *SAE Technical Paper Series*, SAE International400 Commonwealth Drive, Warrendale, PA, United States, 2012 (SAE Technical Paper Series)

[ABK$^+$21] AGHABALI, I.; BAUMAN, J.; KOLLMEYER, P. J.; WANG, Y.; BILGIN, B.; EMADI, A.: 800-V Electric Vehicle Powertrains: Review and Analysis of Benefits, Challenges, and Future Trends. In: *IEEE Transactions on Transportation Electrification* 7 (2021), Nr. 3, S. 927–948

[ABR$^+$10] AZIB, T.; BETHOUX, O.; REMY, G.; MARCHAND, C.; BERTHELOT, E.: An Innovative Control Strategy of a Single Converter for Hybrid Fuel Cell/Supercapacitor Power Source. In: *IEEE Transactions on Industrial Electronics* 57 (2010), Nr. 12, S. 4024–4031

[ÅEB$^+$04] ÅSBOGÅRD, M.; EDSTRÖM, F.; BRINGHED, J.; LARSSON, M.; HELLGREN, J.: Evaluating potential of vehicle auxiliary system coordination using optimal control. In: *7th International Symposium on Advanced Vehicle Control, Arnhem*, 2004, S. 701–706

[AK14] ALAM, M. K.; KHAN, F. H.: Reliability Analysis and Performance Degradation of a Boost Converter. In: *IEEE Transactions on Industry Applications* 50 (2014), Nr. 6, S. 3986–3994

[ASN00] ALFORD, D.; SACKETT, P.; NELDER, G.: Mass customisation — an automotive perspective. In: *International Journal of Production Economics* 65 (2000), Nr. 1, S. 99–110

[BAJ$^+$13] BRABETZ, L.; AYEB, M.; JILWAN, G.; GRAEBEL, P.; KERNER, T.: A new approach to the test, assessment and optimization of robust electrical distribution systems. In: *SAE International Journal of Materials and Manufacturing* 6 (2013), Nr. 3, S. 382–388

[BAS19] BRABETZ, L.; AYEB, M.; SEBASTIAO, D.: Optimization and Evaluation of 12V/48V Architectures Based on EDS Simulation and Real Drive Cycles. In: *SAE Technical Paper Series*, SAE International400 Commonwealth Drive, Warrendale, PA, United States, 2019 (SAE Technical Paper Series)

[BBD+14] Bourdais, R.; Buisson, J.; Dumur, D.; Guéguen, H.; Moroşan, P.-D.: Distributed MPC Under Coupled Constraints Based on Dantzig-Wolfe Decomposition. Version: 2014. In: Maestre, J. M. (Hrsg.); Negenborn, R. R. (Hrsg.): *Distributed Model Predictive Control Made Easy* Bd. 69. Dordrecht: Springer Netherlands, 2014, S. 101–114

[BBGV15] Bubb, H.; Bengler, K.; Grünen, R. E.; Vollrath, M.: *Automobilergonomie*. Wiesbaden: Springer Fachmedien Wiesbaden, 2015

[BBM00] Benini, L.; Bogliolo, A.; Micheli, G. de: A survey of design techniques for system-level dynamic power management. In: *IEEE Transactions on Very Large Scale Integration (VLSI) Systems* 8 (2000), Nr. 3, S. 299–316

[BBM17] Borrelli, F.; Bemporad, A.; Morari, M.: *Predictive Control for Linear and Hybrid Systems*. Cambridge University Press, 2017

[BFB+15] Buechel, M.; Frtunikj, J.; Becker, K.; Sommer, S.; Buckl, C.; Armbruster, M.; Marek, A.; Zirkler, A.; Klein, C.; Knoll, A.: An Automated Electric Vehicle Prototype Showing New Trends in Automotive Architectures. In: *2015 IEEE 18th International Conference on Intelligent Transportation Systems*, IEEE, 2015, S. 1274–1279

[BFMB12] Barthels, A.; Fröschl, J.; Michel, H.-U.; Baumgarten, U.: An Architecture for Power Management in Automotive Systems. Version: 2012. In: Herkersdorf, A. (Hrsg.); Römer, K. (Hrsg.); Brinkschulte, U. (Hrsg.): *Architecture of Computing Systems – ARCS 2012* Bd. 7179. Berlin, Heidelberg: Springer Berlin Heidelberg, 2012, S. 63–73

[BLOT19] Benckendorff, T.; Lapp, A.; Oexner, T.; Thiel, T.: Comparing current and future E/EArchitecture trends of commercial vehicles and passenger cars. In: *19. Internationales Stuttgarter Symposium: Automobil-und Motorentechnik*, 2019, S. 1190–1200

[Büc08] Büchner, S.: *Energiemanagement-Strategien für elektrische Energiebordnetze in Kraftfahrzeugen*. Cuvillier Verlag, 2008

[BW09] Beher, U.; Werthschulte, K.: Energy Management as Configurable System Software Function. In: *SAE Technical Paper Series*, SAE International400 Commonwealth Drive, Warrendale, PA, United States, 2009 (SAE Technical Paper Series)

[CEH18] Caspar, M.; Eiler, T.; Hohmann, S.: Systematic Comparison of Active Balancing: A Model-Based Quantitative Analysis. In: *IEEE Transactions on Vehicular Technology* 67 (2018), Nr. 2, S. 920–934

[CKW15] Chen, H.; Kessels, J.; Weiland, S.: Online adaptive approach for a game-theoretic strategy for Complete Vehicle Energy Management. In: *2015 European Control Conference (ECC)*, IEEE, 2015, S. 135–141

[CPS+15] Cupelli, M.; Ponci, F.; Sulligoi, G.; Vicenzutti, A.; Edrington, C. S.; El-Mezyani, T.; Monti, A.: Power Flow Control and Network Stability in an All-Electric Ship. In: *Proceedings of the IEEE* 103 (2015), Nr. 12, S. 2355–2380

[CSLL13] Christofides, P. D.; Scattolini, R.; La Muñoz de Peña, D.; Liu, J.: Distributed model predictive control: A tutorial review and future research directions. In: *Computers & Chemical Engineering* 51 (2013), S. 21–41

[DCK21] Doan, H. T.; Cho, J.; Kim, D.: Peer-to-Peer Energy Trading in Smart Grid Through Blockchain: A Double Auction-Based Game Theoretic Approach. In: *IEEE Access* 9 (2021), S. 49206–49218

[DLCL21] Deng, R.; Liu, Y.; Chen, W.; Liang, H.: A Survey on Electric Buses—Energy Storage, Power Management, and Charging Scheduling. In: *IEEE Transactions on Intelligent Transportation Systems* 22 (2021), Nr. 1, S. 9–22

[Dör16] Dörfel, R. U.: *Power-Management in Automotiven Systemen - Integration und Umsetzung am Beispiel der PLASA-Plattform*, Technische Universität München, Diss., 2016

[EGLE18] Ehsani, M.; Gao, Y.; Longo, S.; Ebrahimi, K.: *Modern electric, hybrid electric, and fuel cell vehicles*. CRC press, 2018

[ELR08] Emadi, A.; Lee, Y. J.; Rajashekara, K.: Power Electronics and Motor Drives in Electric, Hybrid Electric, and Plug-In Hybrid Electric Vehicles. In: *IEEE Transactions on Industrial Electronics* 55 (2008), Nr. 6, S. 2237–2245

[Ema05] Emadi, A.: *Electrical and computer engineering*. Bd. 125: *Handbook of automotive power electronics and motor drives*. Boca Raton: Taylor & Francis, 2005

[ERWL05] Emadi, A.; Rajashekara, K.; Williamson, S. S.; Lukic, S. M.: Topological Overview of Hybrid Electric and Fuel Cell Vehicular Power System Architectures and Configurations. In: *IEEE Transactions on Vehicular Technology* 54 (2005), Nr. 3, S. 763–770

[ESG04] Ehsani, M.; Shidore, N.; Gao, Y.: On board power management. In: *Power Electronics in Transportation (IEEE Cat. No. 04TH8756)*, 2004, S. 11–17

[EWK06] Emadi, A.; Williamson, S. S.; Khaligh, A.: Power electronics intensive solutions for advanced electric, hybrid electric, and fuel cell vehicular power systems. In: *IEEE Transactions on power electronics* 21 (2006), Nr. 3, S. 567–577

[FBL70] Feigenbaum, E. A.; Buchanan, B. G.; Lederberg, J.: *On generality and problem solving: A case study using the DENDRAL program*. 1970

[FBPZ15] Feng, X.; Butler-Purry, K. L.; Zourntos, T.: A Multi-Agent System Framework for Real-Time Electric Load Management in MVAC All-Electric Ship Power Systems. In: *IEEE Transactions on Power Systems* 30 (2015), Nr. 3, S. 1327–1336

[FBPZC11] Feng, X.; Butler-Purry, K. L.; Zourntos, T.; Chou, H.-M.: Multi-agent system-based real-time load management for NG IPS ships in high/medium voltage level. In: *2011 IEEE/PES Power Systems Conference and Exposition*, IEEE, 2011, S. 1–8

[Fri60] Friedman, M.: *A Program for Monetary Stability*. Ravenio Books, 1960

[GB20] Grossmann, H.; Böttcher, C.: *Pkw-Klimatisierung*. Berlin, Heidelberg: Springer Berlin Heidelberg, 2020

[GCV19] Gurjer, L.; Chaudhary, P.; Verma, H. K.: Detailed Modelling Procedure for Lithium-ion Battery Using Thevenin Equivalent. In: *2019 IEEE International Conference on Electrical, Computer and Communication Technologies (ICECCT)*, IEEE, 2019, S. 1–6

[Gem15] Gemassmer, T.: *Effiziente und dynamische Drehmomenteinprägung in hoch ausgenutzten Synchronmaschinen mit eingebetteten Magneten*. Karlsruhe, KIT, PhD thesis, 2015

[GFKB22] Gerten, M.; Frei, S.; Kiffmeier, M.; Bettgens, O.: Voltage Stability of Automotive Power Supplies During Tripping Events of Melting and Electronic Fuses. In: *2022 IEEE 95th Vehicular Technology Conference: (VTC2022-Spring)*, IEEE, 2022, S. 1–6

[GFKH09] Gehring, R.; Froschl, J.; Kohler, T. P.; Herzog, H.-G.: Modeling of the automotive 14 V power net for voltage stability analysis. In: *2009 IEEE Vehicle Power and Propulsion Conference*, IEEE, 2009, S. 71–77

[GKO19] Gorelik, K.; Kilic, A.; Obermaisser, R.: Connected Energy Management System for Automated Electric Vehicles With Fail-Operational Powertrain and Powernet. In: *IEEE Transactions on Vehicular Technology* 68 (2019), Nr. 10, S. 9588–9603

[GLD02] Gao, L.; Liu, S.; Dougal, R. A.: Dynamic lithium-ion battery model for system simulation. In: *IEEE Transactions on Components and Packaging Technologies* 25 (2002), Nr. 3, S. 495–505

[GLS+20] Guinea, M.; Litton, I.; Smiroldo, R.; Nitsche, I.; Sax, E.: A Proactive Context-Aware Recommender System for In-Vehicle Use. In: *Proceedings of the 2020 4th International Conference on Vision, Image and Signal Processing*. New York, NY, USA: ACM, 2020, S. 1–8

[Gra04a] Graf, H.-M.: Offene Softwarelösung für das Energiemanagement. In: *ATZ - Automobiltechnische Zeitschrift* 106 (2004), Nr. 1, S. 46–50

[Gra04b] GRAF, H.-M.: *Verfahren zur Steuerung der Energieverteilung in einem Verkehrsmittel und Verkehrsmittel, das dieses Verfahren umsetzt.* 2004

[GS17] GEETHA, A.; SUBRAMANI, C.: A comprehensive review on energy management strategies of hybrid energy storage system for electric vehicles. In: *International Journal of Energy Research* 41 (2017), Nr. 13, S. 1817–1834

[GTL+08] GALDUN, J.; TAKAC, L.; LIGUS, J.; THIRIET, J. M.; SARNOVSKY, J.: Distributed Control Systems Reliability: Consideration of Multi-agent Behavior. In: *2008 6th International Symposium on Applied Machine Intelligence and Informatics*, IEEE, 2008, S. 157–162

[GYZ+10] GU, Z.-M.; YANG, D.-G.; ZHANG, X.-F.; LU, L.; LI, K.-Q.; LIAN, X.-M.: Distributed vehicle body electric/electronic system architecture with central coordination control. In: *Proceedings of the Institution of Mechanical Engineers, Part D: Journal of Automobile Engineering* 224 (2010), Nr. 2, S. 189–199

[Hes12] HESSE, B.: *Wechselwirkung von Fahrzeugdynamik und Kfz-Bordnetz unter Berücksichtigung der Fahrzeugbeherrschbarkeit*, Diss., 2012

[HLZ16] HE, J.; LI, J.; ZHANG, L.: A Distribution-Based Model for Electric/Electronic Architectures of Automotive. Version: 2016. In: *Proceedings of SAE-China Congress 2015: Selected Papers* Bd. 364. Singapore: Springer Singapore, 2016, S. 587–598

[HMS09] HENFRIDSSON, O.; MATHIASSEN, L.; SVAHN, F.: Reconfiguring Modularity : Closing Capability Gaps in Digital Innovation. (2009), Nr. 9(22)

[HMVv13] HANK, P.; MULLER, S.; VERMESAN, O.; VAN DEN KEYBUS, J.: Automotive Ethernet: In-vehicle Networking and Smart Mobility. In: *Design, Automation & Test in Europe Conference & Exhibition (DATE), 2013.* New Jersey: IEEE Conference Publications, 2013, S. 1735–1739

[Hoh10] HOHMANN, M.: *Ein synthetischer Ansatz zur Auslegung von Kfz-Bordnetzen unter Berücksichtigung dynamischer Belastungsvorgänge*, Universitätsbibliothek Ilmenau, Diss., 2010

[HSS+22] HENLE, J.; STOFFEL, M.; SCHINDEWOLF, M.; NAGELE, A.-T.; SAX, E.: Architecture platforms for future vehicles: a comparison of ROS2 and Adaptive AUTOSAR. In: *2022 IEEE 25th International Conference on Intelligent Transportation Systems (ITSC)*, IEEE, 2022, S. 3095–3102

[HWK+17] HUANG, Y.; WANG, H.; KHAJEPOUR, A.; HE, H.; JI, J.: Model predictive control power management strategies for HEVs: A review. In: *Journal of Power Sources* 341 (2017), S. 91–106

[Ise16] Isermann, T.: *ACS | automation of complex power systems*. Bd. 39: *A multi-agent-based component control and energy management system for electric vehicles*. 1. Auflage. Aachen: E.ON Energy Research Center, RWTH Aachen University, 2016

[ISM14] Isermann, T.; Sester, S.; Monti, A.: A Multi-Agent Based Energy Management System for Electric Vehicles. In: *2014 IEEE Vehicle Power and Propulsion Conference (VPPC)*, IEEE, 2014, S. 1–6

[JBD16] Jaguemont, J.; Boulon, L.; Dubé, Y.: A comprehensive review of lithium-ion batteries used in hybrid and electric vehicles at cold temperatures. In: *Applied Energy* 164 (2016), S. 99–114

[Jia19] Jiang, S.: Vehicle E/E Architecture and Its Adaptation to New Technical Trends. In: *SAE Technical Paper Series*, SAE International400 Commonwealth Drive, Warrendale, PA, United States, 2019 (SAE Technical Paper Series)

[Joh02] Johnson, V. H.: Battery performance models in ADVISOR. In: *Journal of Power Sources* 110 (2002), Nr. 2, S. 321–329

[KBH09] Kohler, T. P.; Buecherl, D.; Herzog, H.-G.: Investigation of control strategies for hybrid energy storage systems in hybrid electric vehicles. In: *2009 IEEE Vehicle Power and Propulsion Conference*, IEEE, 2009, S. 1687–1693

[KBK+19] Körner, M.-F.; Bauer, D.; Keller, R.; Rösch, M.; Schlereth, A.; Simon, P.; Bauernhansl, T.; Fridgen, G.; Reinhart, G.: Extending the Automation Pyramid for Industrial Demand Response. In: *Procedia CIRP* 81 (2019), S. 998–1003

[KC16] Kouvaritakis, B.; Cannon, M.: *Model Predictive Control*. Cham: Springer International Publishing, 2016

[KET+11] Kohler, T. P.; Ebentheuer, A. W.; Thanheiser, A.; Buecherl, D.; Herzog, H.-G.; Froeschl, J.: Development of an intelligent cybernetic load control for power distribution management in vehicular power nets. In: *2011 IEEE Vehicle Power and Propulsion Conference*, IEEE, 2011, S. 1–6

[KFB+10] Kohler, T.; Froeschl, J.; Bertram, C.; Buecherl, D.; Herzog, H.-G.: Approach of a Predictive, Cybernetic Power Distribution Management. In: *World Electric Vehicle Journal* 4 (2010), Nr. 1, S. 22–30

[KG11a] Kloetzl, J.; Gerling, D.: An interleaved buck-boost-converter combined with a supercapacitor-storage for the stabilization of automotive power nets. In: *2011 IEEE Vehicle Power and Propulsion Conference*, IEEE, 2011, S. 1–6

[KG11b] Kloetzl, J.; Gerling, D.: Stability in automotive power nets: Definitions, algorithms and experimental validation. In: *Proceedings of the 2011 14th European Conference on Power Electronics and Applications*, 2011, S. 1–8

[KGT19] Koo, K.-s.; Govindarasu, M.; Tian, J.: Event prediction algorithm using neural networks for the power management system of electric vehicles. In: *Applied Soft Computing* 84 (2019), S. 105709

[KHP18] Kugele, S.; Hettler, D.; Peter, J.: Data-Centric Communication and Containerization for Future Automotive Software Architectures. In: *2018 IEEE International Conference on Software Architecture (ICSA)*, IEEE, 2018, S. 65–6509

[Kir19] Kirill Gorelik: *Energy management system for automated driving*, Universität Siegen and Fakultät IV Naturwissenschaftlich-Technische Fakultät, Diss., 2019

[KKJ+07] Kessels, J.; Koot, M.; Jager, B. de; van den Bosch, P.; Aneke, N.; Kok, D. B.: Energy Management for the Electric Powernet in Vehicles With a Conventional Drivetrain. In: *IEEE Transactions on Control Systems Technology* 15 (2007), Nr. 3, S. 494–505

[KKv+21] Kilian, P.; Kohler, A.; van Bergen, P.; Gebauer, C.; Pfeufer, B.; Koller, O.; Bertsche, B.: Principle Guidelines for Safe Power Supply Systems Development. In: *IEEE Access* 9 (2021), S. 107751–107766

[Kle04] Klemperer, P.: *Toulouse lectures in economics*. Bd. v. 1: *Auctions: Theory and practice / Paul Klemperer*. Princeton, N.J. and Woodstock: Princeton University Press, 2004

[KLFE11] Konz, M.; Lemke, N.; Försterling, S.; Eghtessad, M.: *Spezifische Anforderungen an das Heiz-Klimasystem elektromotorisch angetriebener Fahrzeuge*. FAT-Schriftenreihe, 2011

[KOB+17] Kugele, S.; Obergfell, P.; Broy, M.; Creighton, O.; Traub, M.; Hopfensitz, W.: On Service-Orientation for Automotive Software. In: *2017 IEEE International Conference on Software Architecture (ICSA)*, IEEE, 2017, S. 193–202

[Koh14] Kohler, T. P.: *Prädiktives Leistungsmanagement in Fahrzeugbordnetzen*. Springer-Verlag, 2014

[KWOH17] Krause, P. C.; Wasynczuk, O.; O'Connell, T.; Hasan, M.: *Introduction to electric power and drive systems*. John Wiley & Sons, 2017

[KWSK02] Krause, P. C.; Wasynczuk, O.; Sudhoff, S. D.; Krause, P. C. A. o. e. m.: *Analysis of electric machinery and drive systems*. 2nd ed. / Paul C. Krause, Oleg Wasynczuk, Scott D. Sudhoff. New York and Great Britain: Wiley-Interscience, 2002 (IEEE Press series on power engineering)

[KWT+10] Kohler, T. P.; Wagner, T.; Thanheiser, A.; Bertram, C.; Buecherl, D.; Herzog, H.-G.; Froeschl, J.: Experimental investigation on voltage stability in vehicle power nets for power distribution management. In: *2010 IEEE Vehicle Power and Propulsion Conference*, IEEE, 2010, S. 1–6

[KYZ+11] Kong, W.; Yang, D.; Zhang, T.; Li, B.; Lian, X.: Vehicle electrical power management system based on an adaptive dual-path controller power supply. In: *2011 International Conference on Electronics, Communications and Control (ICECC)*, IEEE, 2011, S. 2383–2386

[LH02] Leen, G.; Heffernan, D.: Expanding automotive electronic systems. In: *Computer* 35 (2002), Nr. 1, S. 88–93

[LLDL11] Liu, W.; Lian, Z.; Deng, Q.; Liu, Y.: Evaluation of calculation methods of mean skin temperature for use in thermal comfort study. In: *Building and Environment* 46 (2011), Nr. 2, S. 478–488

[LLRB+14] Liebl, J.; Lederer, M.; Rohde-Brandenburger, K.; Biermann, J.-W.; Roth, M.; Schäfer, H.: *Energiemanagement im Kraftfahrzeug*. Springer, 2014

[LM05] Li, B.; Ma, Y.: An Auction-based Negotiation Model in Intelligent Multi-agent System. In: *2005 International Conference on Neural Networks and Brain*, IEEE, 2005, S. 178–182

[LR05] Luk; Rosario: Towards a Negotiation-Based Multi-Agent Power Management System for Electric Vehicles. In: *2005 International Conference on Machine Learning and Cybernetics*, IEEE, 2005, S. 410–416

[LVH11] Lim, H.-T.; Völker, L.; Herrscher, D.: Challenges in a future IP/ethernet-based in-car network for real-time applications. In: Stok, L. (Hrsg.); Dutt, N. (Hrsg.); Hassoun, S. (Hrsg.): *Proceedings of the 48th Design Automation Conference*. New York, NY, USA: ACM, 2011, S. 7–12

[LZ16] Li, S.; Zheng, Y.: *Distributed model predictive control for plant-wide systems*. John Wiley & Sons, 2016

[Mau23] Maurer, J.: *Transactive Control of Coupled Electric Power and District Heating Networks*, Diss., 2023

[MBC11] Maurer, L.; Barroso, L. A.; Chang, J. M.: *Electricity auctions: An overview of efficient practices*. Washington D.C.: World Bank, 2011

[MBH+02] Markel, T.; Brooker, A.; Hendricks, T.; Johnson, V.; Kelly, K.; Kramer, B.; O'Keefe, M.; Sprik, S.; Wipke, K.: ADVISOR: a systems analysis tool for advanced vehicle modeling. In: *Journal of Power Sources* 110 (2002), Nr. 2, S. 255–266

[MDSH14] Ma, J.; Deng, J.; Song, L.; Han, Z.: Incentive Mechanism for Demand Side Management in Smart Grid Using Auction. In: *IEEE Transactions on Smart Grid* 5 (2014), Nr. 3, S. 1379–1388

[MLCA11] Maestre, J. M.; La Muñoz de Peña, D.; Camacho, E. F.; Alamo, T.: Distributed model predictive control based on agent negotiation. In: *Journal of Process Control* 21 (2011), Nr. 5, S. 685–697

[MN14] Maestre, J. M. (Hrsg.); Negenborn, R. R. (Hrsg.): *Distributed Model Predictive Control Made Easy*. Dordrecht: Springer Netherlands, 2014 (Intelligent Systems, Control and Automation: Science and Engineering)

[MOG+09] Marano, V.; Onori, S.; Guezennec, Y.; Rizzoni, G.; Madella, N.: Lithium-ion batteries life estimation for plug-in hybrid electric vehicles. In: *2009 IEEE Vehicle Power and Propulsion Conference*, IEEE, 2009, S. 536–543

[MPPRM+20] Mantilla-Perez, P.; Perez-Rua, J.-A.; Millan, M. A. D.; Dominguez, X.; Arboleya, P.: Power Flow Simulation in the Product Development Process of Modern Vehicular DC Distribution Systems. In: *IEEE Transactions on Vehicular Technology* 69 (2020), Nr. 5, S. 5025–5040

[MRWJ+10] Mohsenian-Rad, A.-H.; Wong, V. W. S.; Jatskevich, J.; Schober, R.; Leon-Garcia, A.: Autonomous Demand-Side Management Based on Game-Theoretic Energy Consumption Scheduling for the Future Smart Grid. In: *IEEE Transactions on Smart Grid* 1 (2010), Nr. 3, S. 320–331

[Nea20] Neacsu, D. O.: *Automotive power systems*. 1st. Boca Raton: CRC press, 2020

[NKK+03] Nuijten, E.; Koot, M.; Kessels, J.; Jager, B. de; Heemels, M.; Hendrix, W.; van den Bosch, P.: Advanced energy management strategies for vehicle power nets. In: *Proc. EAEC 9th Int. Congress: Europ. Automotive Industry Driving Global Changes*, 2003

[NZ20] Nojavan, S.; Zare, K.: *Electricity Markets: New Players and Pricing Uncertainties*. Springer International Publishing, 2020

[ORBE20] Ostadian, R.; Ramoul, J.; Biswas, A.; Emadi, A.: Intelligent Energy Management Systems for Electrified Vehicles: Current Status, Challenges, and Emerging Trends. In: *IEEE Open Journal of Vehicular Technology* 1 (2020), S. 279–295

[PD11] Palensky, P.; Dietrich, D.: Demand Side Management: Demand Response, Intelligent Energy Systems, and Smart Loads. In: *IEEE Transactions on Industrial Informatics* 7 (2011), Nr. 3, S. 381–388

[PPB+07] Polenov, D.; Probstle, H.; Brosse, A.; Domorazek, G.; Lutz, J.: Integration of supercapacitors as transient energy buffer in automotive power nets. In: *2007 European Conference on Power Electronics and Applications*, IEEE, 2007, S. 1–10

[PTS+19] Petersen, P.; Thorgeirsson, A.; Scheubner, S.; Otten, S.; Gauterin, F.; Sax, E.: Training and Validation Methodology for Range Estimation Algorithms. In: *Proceedings of the 5th International Conference on Vehicle Technology and Intelligent Transport Systems*, SCITEPRESS - Science and Technology Publications, 2019, S. 434–443

[RBW+12] Ruf, F.; Barthels, A.; Walla, G.; Winter, M.; Kohler, T. P.; Michel, H.-U.; Froeschl, J.; Herzog, H.-G.: Autonomous load shutdown mechanism as a voltage stabilization method in automotive power nets. In: *2012 IEEE Vehicle Power and Propulsion Conference*, IEEE, 2012, S. 1261–1265

[RDKW19] Romijn, T. C. J.; Donkers, M. C. F.; Kessels, J. T. B. A.; Weiland, S.: A Distributed Optimization Approach for Complete Vehicle Energy Management. In: *IEEE Transactions on Control Systems Technology* 27 (2019), Nr. 3, S. 964–980

[Rei11] Reif, K.: *Bosch Autoelektrik und Autoelektronik*. Wiesbaden: Vieweg+Teubner, 2011

[Rei14] Reif, K.: *Fundamentals of Automotive and Engine Technology*. Wiesbaden: Springer Fachmedien Wiesbaden, 2014

[REL05] Rosario, L. C.; Economou, J. T.; Luk, P.: Multi - Agent Load Power Segregation for Electric Vehicles. In: *2005 IEEE Vehicle Power and Propulsion Conference*, IEEE, 2005, S. 91–96

[RGKS20] Rumez, M.; Grimm, D.; Kriesten, R.; Sax, E.: An Overview of Automotive Service-Oriented Architectures and Implications for Security Countermeasures. In: *IEEE Access* 8 (2020), S. 221852–221870

[RL07] Rosario, L.; Luk, P. C. K.: Applying Management Methodology to Electric Vehicles with Multiple Energy Storage Systems. In: *2007 International Conference on Machine Learning and Cybernetics*, IEEE, 2007, S. 4223–4230

[RLCL14] Rezvanizaniani, S. M.; Liu, Z.; Chen, Y.; Lee, J.: Review and recent advances in battery health monitoring and prognostics technologies for electric vehicle (EV) safety and mobility. In: *Journal of Power Sources* 256 (2014), S. 110–124

[RLEW06] Rosario, L.; Luk, P.; Economou, J. T.; White, B. A.: A Modular Power and Energy Management Structure for Dual-Energy Source Electric Vehicles. In: *2006 IEEE Vehicle Power and Propulsion Conference*, IEEE, 2006, S. 1–6

[Ruf15] RUF, F.: *Auslegung und Topologieoptimierung von spannungsstabilen Energiebordnetzen*, Technische Universität München, Diss., 2015

[RWKT13] RUEGER, J.-J.; WERNET, A.; KECECI, H.-F.; THIEL, T.: MDG1: The New, Scalable, and Powerful ECU Platform from Bosch. Version: 2013. In: *Proceedings of the FISITA 2012 World Automotive Congress* Bd. 194. Berlin, Heidelberg: Springer Berlin Heidelberg, 2013, S. 417–425

[RWM+13] RUF, F.; WINTER, M.; MICHEL, H.-U.; FROESCHL, J.; HERZOG, H.-G.: Experimental Investigations on an Autonomous Load Shutdown Mechanism in Respect to Voltage Stability in Automotive Power Nets. In: *2013 IEEE Vehicle Power and Propulsion Conference (VPPC)*, IEEE, 2013, S. 1–4

[SA03] S. M. LUKIC; A. EMADI: Effects of Electrical Loads on 42V Automotive Power Systems. In: *SAE Transactions* 112 (2003), S. 745–752

[Sca09] SCATTOLINI, R.: Architectures for distributed and hierarchical Model Predictive Control – A review. In: *Journal of Process Control* 19 (2009), Nr. 5, S. 723–731

[SGLS22] SCHINDEWOLF, M.; GRIMM, D.; LINGOR, C.; SAX, E.: Toward a Resilient Automotive Service-Oriented Architecture by using Dynamic Orchestration. In: *2022 IEEE 1st International Conference on Cognitive Mobility (CogMob)*, IEEE, 2022, S. 000147–000154

[SKG17] SHEN, T.; KILIC, A.; GORELIK, K.: Dimensioning of power net for automated driving. In: *Proceedings of EVS30 Electric Vehicle Symposium and Exhibition*, 2017

[SKSS10] SCHMUTZLER, C.; KRUGER, A.; SCHUSTER, F.; SIMONS, M.: Energy efficiency in automotive networks: Assessment and concepts. In: *2010 International Conference on High Performance Computing & Simulation*, IEEE, 2010, S. 232–240

[SSFK16] STEURICH, B.; SCHEIBERT, K.; FREIWALD, A.; KLIMKE, M.: Feasibility Study for a Secure and Seamless Integration of Over the Air Software Update Capability in an Advanced Board Net Architecture. In: *SAE Technical Paper Series*, SAE International400 Commonwealth Drive, Warrendale, PA, United States, 2016 (SAE Technical Paper Series)

[SSG+22] SCHINDEWOLF, M.; STOLL, H.; GUISSOUMA, H.; PUDER, A.; SAX, E.; VETTER, A.; RUMEZ, M.; HENLE, J.: A Comparison of Architecture Paradigms for Dynamic Reconfigurable Automotive Networks. In: *2022 International Conference on Connected Vehicle and Expo (ICCVE)*, IEEE, 2022, S. 1–7

[Ste14] STEINBERG, P.: *Wärmemanagement des Kraftfahrzeugs IX: Energiemanagement.* expert-Verlag, 2014

[TCY+16] Tushar, W.; Chai, B.; Yuen, C.; Huang, S.; Smith, D. B.; Poor, H. V.; Yang, Z.: Energy Storage Sharing in Smart Grid: A Modified Auction-Based Approach. In: *IEEE Transactions on Smart Grid* 7 (2016), Nr. 3, S. 1462–1475

[TGH+15] Tuohy, S.; Glavin, M.; Hughes, C.; Jones, E.; Trivedi, M.; Kilmartin, L.: Intra-Vehicle Networks: A Review. In: *IEEE Transactions on Intelligent Transportation Systems* 16 (2015), Nr. 2, S. 534–545

[THVY16] Tytelmaier, K.; Husev, O.; Veligorskyi, O.; Yershov, R.: A review of non-isolated bidirectional dc-dc converters for energy storage systems. In: *2016 II International Young Scientists Forum on Applied Physics and Engineering (YSF)*, IEEE, 2016, S. 22–28

[TPFH21] Tippe, L.; Pilgrim, L.; Froschl, J.; Herzog, H.-G.: Modular Simulation of Zonal Architectures and Ring Topologies for Automotive Power Nets. In: *2021 IEEE Vehicle Power and Propulsion Conference (VPPC)*, IEEE, 2021, S. 1–5

[TPLS12] Terörde, M.; Purellku, I.; Lücken, A.; Schulz, D.: Modern concepts for the electrical power generation and distribution system on board aircraft. In: *6th International Ege Energy Symposium & Exhibition, Izmir, Turkey*, 2012

[TT12] Tie, S. F.; Tan, C. W.: A review of power and energy management strategies in electric vehicles. In: *2012 4th International Conference on Intelligent and Advanced Systems (ICIAS2012)*, IEEE, 2012, S. 412–417

[TTFH18] Tippe, L.; Taube, J.; Froeschl, J.; Herzog, H.-G.: Introduction of Ring Structures in Future Car Generations' Electrical Systems. In: *2018 IEEE International Conference on Electrical Systems for Aircraft, Railway, Ship Propulsion and Road Vehicles & International Transportation Electrification Conference (ESARS-ITEC)*, IEEE, 2018, S. 1–5

[TTFH19] Taube, J.; Tippe, L.; Froschl, J.; Herzog, H.-G.: Upgradeable Power Net and Its Energy Management Structure. In: *2019 IEEE Vehicle Power and Propulsion Conference (VPPC)*, IEEE, 2019, S. 1–6

[TVE+22] Tippe, L.; Vergara Oberloher, A. de; Ebnicher, M.; Froschl, J.; Herzog, H.-G.: Design and Implementation of a Novel Multi-Domain Management for Automotive Power Nets. In: *2022 IEEE Transportation Electrification Conference & Expo (ITEC)*, IEEE, 2022, S. 147–154

[VA15] Vatanparvar, K.; Al Faruque, M. A.: Battery lifetime-aware automotive climate control for electric vehicles. In: *Proceedings of the 52nd Annual Design Automation Conference*. New York, NY, USA: ACM, 2015, S. 1–6

[VGM+10] Vasquez, J.; Guerrero, J.; Miret, J.; Castilla, M.; Garcia de Vicuna, L.: Hierarchical Control of Intelligent Microgrids. In: *IEEE Industrial Electronics Magazine* 4 (2010), Nr. 4, S. 23–29

[WA08] WALLIN, P.; AXELSSON, J.: A Case Study of Issues Related to Automotive E/E System Architecture Development. In: *15th Annual IEEE International Conference and Workshop on the Engineering of Computer Based Systems (ecbs 2008)*, IEEE, 2008, S. 87–95

[WB18] WATZENIG, D. (Hrsg.); BRANDSTÄTTER, B. (Hrsg.): *Comprehensive Energy Management - Safe Adaptation, Predictive Control and Thermal Management*. Cham: Springer International Publishing, 2018 (SpringerBriefs in Applied Sciences and Technology)

[WFF+16] WINTER, M.; FETTKE, S.; FROESCHL, J.; TAUBE, J.; HERZOG, H.-G.: Using the Viable System Model to control a system of distributed DC/DC converters. In: *2016 IEEE International Conference on Systems, Man, and Cybernetics (SMC)*, IEEE, 2016, S. 002768–002773

[WG20] WANG, D.; GANESAN, S.: Automotive Domain Controller. In: *2020 International Conference on Computing and Information Technology (ICCIT-1441)*, IEEE, 2020, S. 1–5

[WSD+18] WEISS, G.; SCHLEISS, P.; DRABEK, C.; RUIZ, A.; RADERMACHER, A.: Safe adaptation for reliable and energy-efficient E/E architectures. In: *Comprehensive Energy Management-Safe Adaptation, Predictive Control and Thermal Management* (2018), S. 1–18

[WTH18] WINTER, M.; TAUBE, J.; HERZOG, H.-G.: Simulation-based optimization of an energy management for automotive power nets. In: *2018 IEEE International Conference on Electrical Systems for Aircraft, Railway, Ship Propulsion and Road Vehicles & International Transportation Electrification Conference (ESARS-ITEC)*, IEEE, 2018, S. 1–5

[WWWMM01] WELLMAN, M. P.; WALSH, W. E.; WURMAN, P. R.; MACKIE-MASON, J. K.: Auction Protocols for Decentralized Scheduling. In: *Games and Economic Behavior* 35 (2001), Nr. 1-2, S. 271–303

[YKLL16] YANG, D.; KONG, W.; LI, B.; LIAN, X.: Intelligent vehicle electrical power supply system with central coordinated protection. In: *Chinese Journal of Mechanical Engineering* 29 (2016), Nr. 4, S. 781–791

[ZZL+21] ZHU, H.; ZHOU, W.; LI, Z.; LI, L.; HUANG, T.: Requirements-Driven Automotive Electrical/Electronic Architecture: A Survey and Prospective Trends. In: *IEEE Access* 9 (2021), S. 100096–100112

Own Publications and Conference Contributions

[CSAH23] Caspar, M.; Schürmann, T.; Anneken, M.; Hohmann, S.: Active balancing control for distributed battery systems based on cooperative game theory. In: *Journal of Energy Storage* 68 (2023), S. 107585

[GSHH18] Gellrich, T.; Schürmann, T.; Hobus, F.; Hohmann, S.: Model-Based Heater Control Design for Loop Heat Pipes. In: *2018 IEEE Conference on Control Technology and Applications (CCTA)*, IEEE, 2018, S. 527–532

[MLS+23] Majer, N.; Luithle, L.; Schürmann, T.; Schwab, S.; Hohmann, S.: Game-Theoretic Trajectory Planning of Mobile Robots in Unstructured Intersection Scenarios. In: *IFAC-PapersOnLine* 56 (2023), Nr. 2, S. 11808–11814

[RGS+22] Rudolf, T.; Gao, M.; Schürmann, T.; Schwab, S.; Hohmann, S.: Fuzzy Action-Masked Reinforcement Learning Behavior Planning for Highly Automated Driving. In: *2022 8th International Conference on Control, Automation and Robotics (ICCAR)*, IEEE, 2022, S. 264–270

[RSS+22] Rudolf, T.; Schürmann, T.; Skull, M.; Schwab, S.; Hohmann, S.: Data-Driven Automotive Development: Federated Reinforcement Learning for Calibration and Control. Version: 2022. In: Bargende, M. (Hrsg.); Reuss, H.-C. (Hrsg.); Wagner, A. (Hrsg.): *22. Internationales Stuttgarter Symposium*. Wiesbaden: Springer Fachmedien Wiesbaden, 2022 (Proceedings), S. 369–384

[RSSH21] Rudolf, T.; Schürmann, T.; Schwab, S.; Hohmann, S.: Toward Holistic Energy Management Strategies for Fuel Cell Hybrid Electric Vehicles in Heavy-Duty Applications. In: *Proceedings of the IEEE* 109 (2021), Nr. 6, S. 1094–1114

[SGH24] Schürmann, T.; Grobbel, M.; Hohmann, S.: Flexible Automotive Power Management: Formalization of an Auction-based Approach. In: *2024 IEEE Vehicle Power and Propulsion Conference (VPPC)*. Washington, DC, USA: IEEE, Oktober 2024, 1–6

[SKSH22] Schürmann, T.; Kutter, N.; Schwab, S.; Hohmann, S.: Mitigating Power Peaks in Automotive Power Networks by Exploitation of Flexible Loads. In: *2022 XXVIII International Conference on Information, Communication and Automation Technologies (ICAT)*, IEEE, 2022, S. 1–6

[SRSH21] Schürmann, T.; Rafles, A.; Schwab, S.; Hohmann, S.: Distributed and Modular Automotive Power Network Management Based on Auction Theory. In: *2021 IEEE International Conference on Systems, Man, and Cybernetics (SMC)*, IEEE, 2021, S. 2242–2249

[SSH21] Schürmann, T.; Schwab, S.; Hohmann, S.: Application of an Auction based Automotive Power Network Management. In: *FDIBA Conference 2021* Bd. 11. 2021, S. 29–32

[SSH22] Schürmann, T.; Schwab, S.; Hohmann, S.: Communication Scheme for a Modular Automotive Power Network Management. In: *2022 26th International Conference on System Theory, Control and Computing (ICSTCC)*, 2022, S. 582–587

Supervised Theses

[Che18] Chen, Z.: *Modelling and Simulation of the Effects of Different Modulation Methods for Inverter in the Automotive Power Network*. Karlsruhe, Faculty of Electrical Engineering, Karlsruhe Institute of Technology, Master thesis, 2018

[Ehe17] Eherler, S.: *Modellierung einer permanenterregten Synchronmaschine*. Karlsruhe, University of Applied Sciences, Bachelor thesis, 2017

[Fab17] Faber, F.: *Modellierung und Simulation eines DC/DC-Wandlers*. Karlsruhe, Faculty of Electrical Engineering, Karlsruhe Institute of Technology, Bachelor thesis, 2017

[Kie18] Kiefer, P.: *Virtuelle Integration der Bordnetzmanagementfunktionen unter Berücksichtigung der Fahrzeug E/E-Architektur*. Karlsruhe, Faculty of Electrical Engineering, Karlsruhe Institute of Technology, Master thesis, 2018

[Kut20] Kutter, N.: *Umsetzung einer zentralen, modellprädiktiven Regelung für ein Fahrzeugbordnetz*. Karlsruhe, Faculty of Electrical Engineering, Karlsruhe Institute of Technology, Bachelor thesis, 2020

[Lef18] Lefringhausen, R.: *Modellierung und Simulation von Starter-Generatoren im Fahrzeug-Bordnetz*. Karlsruhe, Faculty of Electrical Engineering, Karlsruhe Institute of Technology, Bachelor thesis, 2018

[Möl18] Möller, J.: *Modellierung und Simulation eines elektrischen Kältemittelverdichters*. Karlsruhe, Faculty of Electrical Engineering, Karlsruhe Institute of Technology, Bachelor thesis, 2018

[Raf20] Rafles, A.: *Conception and Implementation of Market-Based Methods of the Demand-Side-Management for an Automotive Power Network*. Karlsruhe, Faculty of Electrical Engineering, Karlsruhe Institute of Technology, Bachelor thesis, 2020

[Tis17] Tissen, R.: *Modellierung und Simulation eines Pkw-Bordnetzes*. Karlsruhe, Faculty of Electrical Engineering, Karlsruhe Institute of Technology, Bachelor thesis, 2017

[Wu18] Wu, L.: *Erweiterung und Validierung der Bordnetzsimulation eines Hybridfahrzeugs*. Karlsruhe, Faculty of Mechanical Engineering, Karlsruhe Institute of Technology, Master thesis, 2018

[Yin18] Yin, S.: *Modeling of Components of the Automotive Power Network with Blackbox and Greybox Models and Parameter Identification*. Karlsruhe, Faculty of Mechanical Engineering, Karlsruhe Institute of Technology, Master thesis, 2018

[Yin21] Yin, Y.: *A Negotiation-based Distributed Model Predictive Control for Modular and Flexible Power Management in Automotive Power Networks*. Karlsruhe, Faculty of Electrical Engineering, Karlsruhe Institute of Technology, Master thesis, 2021

Karlsruher Beiträge zur Regelungs- und Steuerungstechnik (ISSN 2511-6312)
Institut für Regelungs- und Steuerungssysteme

Band 01 Diehm, Gunter
Identifikation des menschlichen Bewegungsverhaltens auf der Basis von Primitiven.
ISBN 978-3-7315-0608-9

Band 02 Flad, Michael
Kooperative Regelungskonzepte auf Basis der Spieltheorie und deren Anwendung auf Fahrerassistenzsysteme.
ISBN 978-3-7315-0610-2

Band 03 Eckert, Marius
Modellbasierte Identifikation fraktionaler Systeme und ihre Anwendung auf die Lithium-Ionen-Zelle.
ISBN 978-3-7315-0690-4

Band 04 Krebs, Stefan
Intervallbeobachter für lineare parametervariante Systeme und deren Anwendung auf die Asynchronmaschine.
ISBN 978-3-7315-0857-1

Band 05 Kaspar, Stephan
Fahrdynamikuntersuchungen eines Elektrofahrzeugs mit Einzelrad-Hinterradantrieb.
ISBN 978-3-7315-0916-5

Band 06 Sauter, Patrick S.
Modellierung und zentrale prädiktive Regelung von multimodalen Energieverteilnetzen.
ISBN 978-3-7315-0963-9

Band 07 Kupper, Martin
Verteilte Zustandsschätzung fraktionaler Systeme und ihre Anwendung auf Lithium-Ionen-Batteriesysteme.
ISBN 978-3-7315-0971-4

Band 08 Merkert, Lennart
Optimal Scheduling of Combined Heat and Power Generation Considering Heating Grid Dynamics.
ISBN 978-3-7315-1056-7

Die Bände sind unter www.ksp.kit.edu als PDF frei verfügbar oder als Druckausgabe bestellbar.

Band 09 Ludwig, Julian
Automatisierte kooperative Transition einer Regelungsaufgabe zwischen Mensch und Maschine am Beispiel des hochautomatisierten Fahrens.
ISBN 978-3-7315-1069-7

Band 10 Inga Charaja, Juan Jairo
Inverse Dynamic Game Methods for Identification of Cooperative System Behavior.
ISBN 978-3-7315-1080-2

Band 11 Schnurr, Christoph Xaver
Ein Verfahren zur lexikographischen modellprädiktiven Regelung mit der Anwendung auf eine permanenterregte Synchronmaschine.
ISBN 978-3-7315-1095-6

Band 12 Schwab, Stefan
Guaranteed Verification of Dynamic Systems.
ISBN 978-3-7315-0965-3

Band 13 Pfeifer, Martin
Automated Model Generation and Observer Design for Interconnected Systems: A Port-Hamiltonian Approach.
ISBN 978-3-7315-1135-9

Band 14 König, Alexander
Absicherung hochautomatisierten Fahrens durch passiven virtuellen Dauerlauftest.
ISBN 978-3-7315-1141-0

Band 15 Stark, Oliver
Parameter- und Ordnungsidentifikation von fraktionalen Systemen mit einer Anwendung auf eine Lithium-Ionen-Batteriezelle.
ISBN 978-3-7315-1187-8

Band 16 Köpf, Florian
Adaptive Dynamic Programming: Solltrajektorienfolgeregelung und Konvergenzbedingungen.
ISBN 978-3-7315-1193-9

Die Bände sind unter www.ksp.kit.edu als PDF frei verfügbar oder als Druckausgabe bestellbar.

Band 17 Kölsch, Lukas
Dynamic Incentives for Optimal Control of Competitive Power Systems.
ISBN 978-3-7315-1209-7

Band 18 Schwartz, Manuel
Topologie-Optimierung eines radselektiv angesteuerten Fahrzeugs basierend auf einer optimalen Fahrzeugführungsregelung.
ISBN 978-3-7315-1222-6

Band 19 Rothfuß, Simon
Human-Machine Cooperative Decision Making.
ISBN 978-3-7315-1223-3

Band 20 Gellrich, Thomas Christoph
Dynamical Modeling and Control of Multiphase Heat Transport Systems Based on Loop Heat Pipes.
ISBN 978-3-7315-1231-8

Band 21 Maurer, Jona
Transactive Control of Coupled Electric Power and District Heating Networks.
ISBN 978-3-7315-1276-9

Band 22 Puccetti, Luca
Self-Learning Longitudinal Control for On-Road Vehicles.
ISBN 978-3-7315-1290-5

Band 23 Varga, Bálint
Limited Information Shared Control and its Applications to Large Vehicle Manipulators.
ISBN 978-3-7315-1325-4

Band 24 Walz, Fabian
Modellbasierte prädiktive Längsdynamikregelung für künftige Fahrerassistenz- und Automatisierungssysteme.
ISBN 978-3-7315-1331-5

Band 25 Strehle, Felix
A Framework for Decentralized Stabilization in Networked Energy System: A Passivity-Based Approach
ISBN 978-3-7315-1358-2

Die Bände sind unter www.ksp.kit.edu als PDF frei verfügbar oder als Druckausgabe bestellbar.

Band 26 Esther Sophie Bischoff
Time-Extended Multi-Robot Task Allocation – A Reoptimization Framework with Provable Performance.
ISBN 978-3-7315-1449-7

Band 27 Albertus Johannes Malan
Passivity-Based Stabilisation and Coordination in Networked Multi-Energy Systems.
ISBN 978-3-7315-1453-4

Band 28 Schürmann, Tobias
Flexible Automotive Power Management: A Predictive Auction-Based Approach.
ISBN 978-3-7315-1454-1